BIOTECHNOLOGY IN THE MODERN MEDICINAL SYSTEM

Advances in Gene Therapy, Immunotherapy, and Targeted Drug Delivery

BIOTECHNOLOGY IN THE MODERN MEDICINAL SYSTEM

Advances in Gene Therapy, Immunotherapy, and Targeted Drug Delivery

Edited by

Rajesh K. Kesharwani, PhD

Krishna Misra, PhD

AAP | APPLE ACADEMIC PRESS

First edition published 2022

Apple Academic Press Inc.
1265 Goldenrod Circle, NE,
Palm Bay, FL 32905 USA

4164 Lakeshore Road, Burlington,
ON, L7L 1A4 Canada

CRC Press
6000 Broken Sound Parkway NW,
Suite 300, Boca Raton, FL 33487-2742 USA

2 Park Square, Milton Park,
Abingdon, Oxon, OX14 4RN UK

Library and Archives Canada Cataloguing in Publication

Title: Biotechnology in the modern medicinal system : advances in gene therapy, immunotherapy, and targeted drug delivery / edited by Rajesh K. Kesharwani, PhD, Krishna Mishra, PhD.

Names: Kesharwani, Rajesh Kumar, 1978- editor. | Mishra, Krishna, 1938- editor.

Description: First edition. | Includes bibliographical references and index.

Identifiers: Canadiana (print) 20210123729 | Canadiana (ebook) 20210123796 | ISBN 9781771889728 (hardcover) | ISBN 9781774638248 (softcover) | ISBN 9781003129783 (PDF)

Subjects: LCSH: Gene therapy. | LCSH: Immunotherapy. | LCSH: Drug delivery systems.

Classification: LCC RB155.8 .B56 2021 | DDC 616/.042—dc23

Library of Congress Cataloging-in-Publication Data

Names: Kesharwani, Rajesh K., 1978- editor. | Misra, Krishna, 1938- editor.

Title: Biotechnology in the modern medicinal system : advances in gene therapy, immunotherapy, and targeted drug delivery / edited by Rajesh K. Kesharwani, Krishna Misra.

Description: First edition. | Palm Bay, FL : Apple Academic Press, 2021. | Includes bibliographical references and index. | Summary: "Biotechnology in the Modern Medicinal System: Advances in Gene Therapy, Immunotherapy, and Targeted Drug Delivery presents an informative picture of the state-of-the-art research and development of actionable knowledge in medical biotechnology, specifically involving gene therapy, immunotherapy, and targeted drug delivery systems. The book includes novel approaches for therapy of various ailments and the real-world challenges and complexities of the current drug delivery methodologies and techniques. The volume helps to bridge the gap between academic research and real-time clinical applications and the needs of medical biotechnology methods. This edited book also provides a detailed application of medical biotechnology in drug discovery and the treatment of various deadly diseases. Chapters discuss targeted drug delivery to specific sites to avoid possible entry to non-targeted sites, minimizing adverse effects. The volume provides information about the roles of alternative routes of drug targeting, like intranasal and transdermal, resulting in improving patient compliance. Targeted drug delivery is explored for several health issues, such as neurodegenerative disorders, cancer, malaria, and hemoglobin disorders. Also considered is the role of genes in various genetic diseases and gene therapy, and immunogene therapy as alternative approaches to conventional cancer therapy. Finally, the book investigates the important role of computers in biotechnology to accelerate research and development in the modern medicinal field for better and optimum results. Studies show that significant improvement has been observed in the development of a faster and less invasive diagnostic system for the treatment of diseases by utilizing both artificial intelligence (AI) and biotechnology. This valuable volume provides a wealth of information that will be valuable to scientists and researchers, faculty, and students"-- Provided by publisher.

Identifiers: LCCN 2021003386 (print) | LCCN 2021003387 (ebook) | ISBN 9781771889728 (hardcover) | ISBN 9781774638248 (paperback) | ISBN 9781003129783 (ebook)

Subjects: MESH: Biomedical Technology | Immunotherapy | Genetic Therapy | Drug Delivery Systems

Classification: LCC RM282.I44 (print) | LCC RM282.I44 (ebook) | NLM W 82 | DDC 615.3/7--dc23

LC record available at https://lccn.loc.gov/2021003386

LC ebook record available at https://lccn.loc.gov/2021003387

ISBN: 978-1-77188-972-8 (hbk)
ISBN: 978-1-77463-824-8 (pbk)
ISBN: 978-1-00312-978-3 (ebk)

About the Editors

Rajesh K. Kesharwani, PhD, MTech-IT
Associate Professor, Department of Computer Application, Nehru Gram Bharati (Deemed to be University), Prayagraj, Uttar Pradesh, India

Rajesh K. Kesharwani, PhD, has more than 10 years of research and eight years of teaching experience in various institutes of India, imparting bioinformatics and biotechnology education. He has received several awards, including the NASI-Swarna Jayanti Puruskar by The National Academy of Sciences of India. He has supervised PhD, postgraduate, and undergraduate students for their research work and has authored over 40 peer-reviewed articles, more than 20 book chapters, and over 10 edited books with international publishers. He has been a member of many scientific communities as well as a reviewer for many international journals. He has presented many papers in various national and international conferences.

Dr. Kesharwani received his PhD from the Indian Institute of Information Technology, Allahabad, India, and worked at NIT Warangal. He was a recipient of the Ministry of Human Resource Development (India) Fellowship and Senior Research Fellowship from the Indian Council of Medical Research, India. His research fields of interest are medical informatics, protein structure and function prediction, computer-aided drug designing, structural biology, drug delivery, cancer biology, nanobiotechnology, and biomedical sciences.

Prof (Mrs.) Krishna Misra, PhD, FNASc, FBRSI
(Honorary Professor, Indian Institute of Information Technology-Allahabad, Devghat, Jhalwa, Prayagraj, Uttar Pradesh, India)

Krishna Misra, PhD, is superannuated as Professor of Chemistry at the University of Allahabad, India, where she occupied the Chair of Head, Biochemistry Department and was first coordinator of the Center for Biotechnology. She is at present an honorary professor and was Coordinator of Indo-Russian Center for Biotechnology at IIIT-Allahabad as well as honorary professor at the Centre of Biomedical Research, Lucknow. She is also a senior scientist and fellow of the National Academy of Sciences India, where she was formerly general secretary. In addition, she is chief advisor for India Pesticides Ltd., Lucknow, and was a member of the advisory board of Biotech Park, Lucknow. She was awarded a NASI Platinum Jubilee Senior Scientist Fellowship. She is chairperson of the STEM program of the Department of Science and Technology and Chemistry Advisory Board of Uttar Pradesh Council of Science and Technology. She was an expert on many selection committees and is also on the editorial boards of many national and international journals. She has supervised many PhD students and has published several books, about 20 reviews, a dozen book chapters, and many papers at national and international conferences. She holds Indian and US patents. She has visited Japan as a UNESCO fellow; the UK, sponsored by the British Council; and the USA. She had been awarded many research projects, including three prestigious international projects from USA. She had been member of the task force of biotechnology for central India government and is founder member and fellow of many national and international science societies.

Professor Misra earned her PhD in 1964 from Delhi University under the supervision of late Prof T. R. Seshadri, FRS, and Padma Bhushan. From 1966 to 1999 worked at Allahabad University as lecturer, reader, and professor. She has supervised 55 PhD students, has published over 250 papers, three books, about 20 reviews, and a dozen book chapters. She has also presented papers at about 100 national and international conferences and holds several Indian and U.S. patents. Dr. Misra has visited Japan (as a UNESCO fellow), the U.K. (sponsored by the British Council), and the USA (invited for papers/talks/chair) a number of times to deliver lectures and to participate in scientific discussions. She has been awarded a large number of research projects, including three prestigious international projects from the USA. She has been member of a task force of biotechnology for Central government and is a founding member and fellow of a number of national and international science societies. She has been teaching graduate and postgraduate classes and also conducting research in organic chemistry, biochemistry, biotechnology, bioinformatics, biomedical engineering, and research methodology. Her fields of scientific research have been chemistry of naturally occurring herbal products of biological importance, DNA synthesis, oligonucleotide chemistry, tagging with fluorescent tags, antisense therapy, chemoinformatics, systems biology, computer-aided drug design, molecular medicine, biomedical engineering, targeted drug designing, nanotechnology, and nano-biotechnology. She has delivered lectures all over India and abroad about her research work on women empowerment, climate change, scientific writing, etc.

Contents

Contributors

Sukanya Bhoumik
Department of Biochemistry, University of Allahabad, Allahabad, India

Princy Choudhary
Indian Institute of Information Technology Allahabad, Devghat, Jhalwa, Allahabad 211015, Uttar Pradesh, India

Soma M. Ghorai
Department of Zoology, Hindu College, University of Delhi, Delhi 110007, India

Kapil Kumar Gupta
Department of Computer Science and Engineering, Shri Ramswaroop Memorial University, Barabanki, India

Neha Kapoor
Department of Chemistry, Hindu College, University of Delhi, Delhi 110007, India

Rajesh K. Kesharwani
Department of Advanced Science & Technology, NIET, Nims University Rajasthan, Jaipur, Rajasthan, India, E-mail: rajiiita06@gmail.com

Upendra Kumar
Department of Computer Science and Engineering, Institute of Engineering and Technology, Lucknow 226021, India

Ashutosh Mani
Department of Biotechnology, Motilal Nehru National Institute of Technology Allahabad 211004, India

Pramod Kumar Maurya
Department of Biotechnology, Motilal Nehru National Institute of Technology Allahabad 211004, India

Neha Shree Maurya
Department of Biotechnology, Motilal Nehru National Institute of Technology Allahabad 211004, India

Krishna Misra
Biochemistry Department, University of Allahabad, India, E-mail: krishnamisra@hotmail.com

Vandana Patravale
Department of Pharmaceutical Sciences and Technology, Institute of Chemical Technology, Matunga (E), Mumbai 400019, Maharashtra, India

Sreeranjini Pulakkat
Department of Pharmaceutical Sciences and Technology, Institute of Chemical Technology, Matunga (E), Mumbai 400019, Maharashtra, India

Syed Ibrahim Rizvi
Department of Biochemistry, University of Allahabad, Allahabad, India

Sangeeta Singh
Indian Institute of Information Technology Allahabad, Devghat, Jhalwa, Allahabad 211015, Uttar Pradesh, India

Sarika Wairkar
Shobhaben Pratapbhai Patel School of Pharmacy and Technology Management, SVKM's NMIMS, V.L. Mehta Road, Vile Parle (W), Mumbai 400 056, Maharashtra, India

Viroj Wiwanitkit
Dr DY Patil University, Pune, India

Sora Yasri
KMT Primary Care Center, Bangkok, Thailand

Abbreviations

AAV	adeno-associated virus
ABC	ATP-binding cassette
ACT	adoptive T cell transfer
Ad	adenovirus
ADA	adenosine deaminase
AI	artificial intelligence
ANN	artificial neural network
AMT	adsorptive-mediated transcytosis
APCs	antigen-presenting cells
BBB	blood–brain barrier
BCSFB	blood cerebrospinal fluid barrier
BDNF	brain-derived neurotrophic factor
bFGF	basic fibroblast growth factor
CARs	chimeric antigen receptors
CAR-T	chimeric antigen receptor-T
CED	convection-enhanced delivery
CFTR	cystic fibrosis transmembrane conductance regulator
CLL	chronic lymphocytic leukemia
CNS	central nervous system
CNTF	ciliary neurotrophic factor
CPP	cell-penetrating peptides
CRLBP	cathode ray local binary pattern
CSF	cerebrospinal fluid
DA	dopamine
DC	dendritic cells
DNNs	deep neural networks
ECs	endothelial cells
ELISA	enzyme-linked immunosorbent assay
EPO	erythropoietin
EPR	enhanced permeability and retention
EVAc	ethylene vinyl acetate copolymer
FDA	Food and Drug Administration
FP	heme iron

FUS	focused ultrasound
GA	genetic algorithm
GDNF	glial cell line derived neurotrophic factor
GHSR	growth hormone secretagogue receptor
GFP	green fluorescence protein
GM-CSF	granulocyte-macrophage colony stimulating factor
HaaS	healthcare-as-a-service
HDL	high density lipoproteins
HDR	homology-directed repair
hGDNF	human glial cell line derived neurotrophic factor
IFN	interferon
IoT	Internet of Things
LBP	local binary pattern
LDP	local derivative pattern
LDL	low density lipoproteins
k-NN	k-nearest neighbor algorithms
MAb	monoclonal antibody
MAS	macrophage activation syndrome
MHC	major histocompatibility complex
ML	machine learning
MLTs	machine learning techniques
MoA	mechanism of action
MSX1	msh homeobox 1
NGF	nerve growth factor
NHEJ	nonhomologous end joining
NK	natural killer
NPs	natural products
NTN	neurturin
OL	odorranalectin
PAMs	pharmacologically active microparticles
PBCA	polybutyl cyanoacrylate
PEG–PLGA	polyethylene glycol–polylactic-*co*-glycolic acid
PgP	P-glycoprotein
POMC	pro-opiomelanocortin
RBCs	red blood cells
RMT	receptor-mediated transcytosis
SA	streptavidin
SCID	severe combined immunodeficiency

SHOX	short stature-homeobox
siRNA	small-interfering RNAs
SOMs	self-organizing feature maps
SVMs	support vector machines
TALENS	transcription activator-like effector nucleases
TCI	transcutaneous immunization
TCR	T cell receptor
TfR	transferrin receptors
TLR	toll-like receptor
TP	transportan
UN	urocortin
VBD	vector-borne disease
VLDL	very low density lipoproteins

Foreword

It gives me great pleasure to write the foreword for the book, *Biotechnology in the Modern Medicinal System: Advances in Gene Therapy, Immunotherapy, and Targeted Drug Delivery*. It provides detailed information on the application of biotechnology in gene therapy, immunotherapy, drug delivery systems, and artificial intelligence in medicine.

This book covers medical processes such as, for example, designing of organisms to produce drugs, engineering of genetic cures through genetic manipulation (gene therapy), diagnostics, drug discovery, and targeted delivery. The drug delivery systems (DDS) that use a variety of carriers have been developed in order to minimize drug degradation, to prevent side effects, and to increase drug bioavailability. The DDS that were created for traditional oral and intravenous administration have been expanded to transdermal, nasal, ocular, buccal, intramuscular, rectal, intrauterine, vaginal, and pulmonary administration, ceramic implants, etc. These systems have been well described in this book.

The role of genes in therapeutics, termed as "gene therapy," is emerging as a tool to provide treatment against numerous genetic and related deadly diseases. The authors have provided a of about ongoing research and trials on gene therapy for management of genetic disorders. A large number of genetic disorders are known to arise from either type of gene mutations and are almost lethal or may cause variety of abnormalities. Disease susceptibility of an individual depends on genomic conformation as well as external exposure.

This book provides a wealth of information that will be valuable to scientists, researchers, faculty, and students in the field of biomedical sciences. All the contributing authors demonstrate an exceptional expertise in the field of medicinal biotechnology research and development and provide a global perspective on current and future advances in new drug discovery.

I congratulate the editors for bringing together experts in the field of medical biotechnology and the authors for their excellent contributions.

Prof. C. L. Khetrapal
Distinguished Professor and Founder
Director of Centre of Biomedical Research,
Sanjay Gandhi Post Institute of Medical Sciences Campus,
Lucknow, Uttar Pradesh, India

Preface

This book, *Biotechnology in the Modern Medicinal System: Advances in Gene Therapy, Immunotherapy, and Targeted Drug Delivery,* presents a full picture of the state-of-the-art research and development of actionable knowledge in medical biotechnology involving, specifically, gene therapy, immunotherapy, and targeted drug delivery systems. The book includes novel approaches for therapy of various ailments and the real-world challenges and complexities to the current drug delivery methodologies and techniques.

As is evident from latest discussions at various globally held conferences and seminars, several medical biotechnology methods have been used but only a few of them have been validated in actual practice. A major reason for the above situation, we believe, is the gap between academic research and real-time clinical applications and needs.

The present book includes eight chapters containing information about the role of biotechnology in the modern medicinal system for human welfare. This edited book also provides a detailed application of medical biotechnology in drug discovery and the treatment of various deadly diseases.

Chapter 1, entitled "Role of Naturally Occurring Lead Compounds as Potential Drug Targets Against Malaria," authored by Neha Kapoor and Soma M. Ghorai, explains the diversity of natural products (NPs) obtained from telluric and marine plants as well as from microorganisms in the treatment of variety of communicable and noncommunicable diseases. However, this chapter is restricted to document those lead compounds that are used in the management of malaria, a mosquito-borne disease. A wide range of compounds resulting from natural product sources such as endoperoxide, isonitrile derivatives, alkaloids, and non-alkaloid derivatives (terpenes, flavonoids, quinones, phenols, polyethers, and peptides), have been found to have promise as antimalarials.

Chapter 2, entitled "Targeted Drug Delivery," by Sangeeta Singh and Princy Choudhary, focuses on delivering the drug through drug-carrier complex at a higher concentration to targeted sites, resulting in minimum possible entry to nontargeted sites, minimizing the adverse effects. Administration of drugs to specific cavities, such as the like pleural cavity, peritoneal cavity, cerebral ventricles, or tissues, such as tumors or Kupffer cells of liver and intracellular localization of drugs or drug carrier system or targeting of DNA and proteins to a cell, are all possible via targeted drug delivery. A wide range of carriers is available that are specifically used according to the requirements and nature of the route and the target site.

Author Vandana Patravale and her associates have written Chapter 3, entitled "Targeted Delivery of Biopharmaceuticals for Neurodegenerative Disorders," which provides information about the role of alternative routes of drug targeting, such as intranasal and transdermal, resulting in improving the patient compliance. Similarly, noninvasive techniques have been successfully applied for brain delivery of biopharmaceuticals. In addition, several colloidal carriers have been explored for passive targeting of biopharmaceuticals whereas more precision has been achieved by active targeting with ligands. This chapter summarizes the advanced delivery approaches of biopharmaceuticals to the brain and the preclinical studies associated with them in treating complex neurodegenerative disorders. Nevertheless, a systematic clinical investigation is necessary before exploring their therapeutic translation.

The role of genes in various genetic diseases has been well described in Chapter 4, entitled "Genes in Genetic Disease," by Rizvi and his associate. Genes are the hereditary factors that are passed from generation to generation and are responsible for determining the genotypic as well as phenotypic traits in an individual. Every human being has about 20,000–25,000 genes, which encode for a variety of polypeptides and proteins. A large number of genetic disorders are known that arise from either type of gene mutations and are almost lethal or may cause variety of abnormalities. Disease susceptibility of an individual depends on genomic confirmation as well as external exposure.

Chapter 5, entitled "Current Perspectives and Trends in Gene Therapy and Their Clinical Trials," by Ashutosh Mani and his colleagues, well describes the need for target-specific modifications in human genome for the purpose of treating the genetic diseases. Gene therapy is known as a method of altering and mutating the genes to be used for therapeutic purpose. This approach is broad in an experimental sense and still needs various new developments and is still in its experimentally driven phase. This chapter focuses on the trends that are being followed up by the researchers now a days for current gene therapy-based clinical trials.

In the current drug discovery scenario, immunogenes are playing a very important role, and it is well documented in Chapter 6, entitled "Immunogene Therapy in Cancer," by Pulakkat and Patravale, in a precise manner. Immunotherapy has long been investigated as a potent, alternative approach to conventional cancer therapy; however, the clinical translations are limited. Immuno-oncology and genomics have experienced vast advancements in the recent past, and this proliferation of knowledge gave birth to the field of cancer immunogene therapy. The different approaches explored in immunogene therapy including ex vivo manipulation of T-lymphocytes, transferring immunostimulatory genes to tumor cells and antigen-presenting cells, in vivo genetic modulation using viral and nonviral vectors etc. have been discussed in this chapter.

The role of genes in therapeutics, nowadays termed as "gene therapy" has a very important role against various diseases, including hemoglobin disorders, as described in Chapter 7, entitled "Gene Therapy for Hemoglobin Disorders," written by Sona Yasri and Viroj Wiwanitkit. The chapter describes how gene therapy is emerging as a tool to provide treatment against numerous genetic and related deadly diseases. In the present chapter, the authors provide a glimpse of ongoing research and trials on gene therapy for management of hemoglobin disorders. Several in silico and in vitro studies prove that gene therapy is useful for management of hemoglobin disorders.

The role of computers in biotechnology accelerates the research and development specifically in modern medicinal field with better and optimum results. Nowadays artificial intelligence is playing a very pivotal role in many fields. Chapter 8, entitled "Artificial Intelligence and Biotechnology: The Golden Age of Medical Research," by Upendra Kumar and Kapil Kumar, describes the significant improvement that has been observed in the development of a faster and less invasive diagnostic system for the treatment of diseases by utilizing both artificial intelligence (AI) and biotechnology. The biomedical research landscape is changing and evolving.

The present book, entitled *Biotechnology in the Modern Medicinal System: Advances in Gene Therapy, Immunotherapy, and Targeted Drug Delivery*, provides detailed information on the application of biotechnology in gene therapy, immunotherapy, drug delivery systems, and artificial intelligence in medicine. This valuable volume provides a wealth of information that will be beneficial to scientists and researchers, faculty, and students.

CHAPTER 1

Role of Naturally Occurring Lead Compounds as Potential Drug Targets against Malaria

NEHA KAPOOR[1*] and SOMA M. GHORAI[2]

[1]*Department of Chemistry, Hindu College, University of Delhi, Delhi 110007, India*

[2]*Department of Zoology, Hindu College, University of Delhi, Delhi 110007, India*

[*]*Corresponding author. E-mail: nehakapoor@hindu.du.ac.in*

ABSTRACT

This chapter overviews the diversity of natural products (NPs) obtained from telluric plants, marine-, and microorganisms in the treatment of variety of communicable and noncommunicable diseases. A historical perspective is being given to the readers about the discovery and use of various NPs as lead compounds for the synthesis of various drugs. Screening of natural resources to generate new lead compounds has been possible due to their enormous structural diversity and medicinal significance. Generation of many lead compounds with different structural analogs having fewer side effects and more pharmacological activity can be obtained by molecular modifications of their functional groups. Among others, much has been acknowledged about lead compounds that hold the promise as potential drug targets against almost all the vector-borne diseases. However, this chapter is restricted to document those lead compounds that are used in the management of malaria, a mosquito-borne disease. A wide range of compounds resulting from natural product sources such as endoperoxide, isonitrile derivatives, alkaloids, and nonalkaloid derivatives (terpenes, flavinoids, quinones, phenols, polyethers, and peptides) have been found to have promise as antimalarials. Such lead compounds as antimalarials with unique functional groups and chemical backbones holds key to future drug synthesis against *Plasmodium* parasites.

1.1 HISTORICAL PERSPECTIVE

Every era had seen successive development of use of natural products (NPs) to benefit societies and is passed on to another with improvement in its effectiveness through generations. Approximately, 5000 years ago, medicinal plants' were the first candidates to be used and the early evidences have been found on a Sumerian clay slab from Nagpur. The inscriptions mentioned about 250 various plants and 12 recipes for drug preparation from plants like poppy, henbane, and mandrake (Kelly, 2009). The Chinese Emperor Shen Nung was the first to have documented a book called Pen T'Sao, circa 2500 BC, which defines 365 drugs from dried parts of medicinal plants including *Rhei rhisoma*, *Theae folium*, camphor, the great yellow gentian, podophyllum, ginseng, cinnamon bark, jimson weed, and ephedra (Wiart, 2006; Petrovska, 2012). Likewise, ancient Indian Vedas including the Charaka and Samhitas, which dates back to 1000 BC, recognized 341 medicinal plants and 516 drugs of the Indian Ayurvedic system (Dev, 1999; Kapoor, 2017). Many plant extracts of pomegranate, castor oil plant, senna, aloe, onion, fig, willow, coriander, juniper, common centaury, and so on have been mentioned along with garlic, a collection of 700 plant species used for therapeutics in Ebers Papyrus, circa 1550 BC (Tucakov, 1964). Talmud, the holy book of Jews, refers to the use of aromatic plants as incense or myrtle (Dimitrova, 1999). Homer's epics, *The Iliad* and *The Odysseys* (ca. 800 BC), also mentions use of 63 plant species as pharmacotherapeutic agents (Toplak Galle, 2005). Greek scholars such as Herodotus and Pythagoras (500 BC) mentioned about the therapeutic values of garlic, castor oil, mustard, and cabbage. Hippocrates (459–370 BC) categorized nearly 300 medicinal plants conferring to their functional effects. Common centaury (*Centaurium umbellatum* Gilib.) and wormwood were administered against fever; garlic was used against intestine parasites; opium, deadly nightshade, mandrake, and henbane were consumed as narcotics; haselwort and fragrant hellebore were considered emetics; oak and pomegranate as astringents; while sea onion, parsley, celery, garlic, and asparagus were used as diuretics (Bojadzievski, 1992;

Gorunovic and Lukic, 2001). Theophrast (371–287 BC), also known as the "Father of Botany," was given due credit for cataloging more than 500 medicinal plants (Bazala, 1943; Nikolovski, 1995). The era from 23 to 79 AD was distinguished by two important contemporaries in pharmacy, Dioscorides (77 AD) and Pliny the Elder (23–79 AD). Both were credited to have traveled and documented more than 1000 medicinal plants (Tucakov, 1990; Toplak Galle, 2005). Galen (131–200 AD), a Roman physician, familiarized numerous new plants as remedial drugs which Dioscorides had not described, for example, *Uvae ursi folium* is used as an uroantiseptic and a mild diuretic. Some plants were used as insecticides, namely, *Veratrum album*, *Alium sativum*, *Urtica dioica*, *Cucumis sativus*, *Achilea millefolium*, *Lavandula officinalis*, *Artemisia maritime* L., and *Sambuci flos* (Bojadzievski, 1992) and by the seventh century, *Ocimum basilicum*, *Rosmarinus officinalis*, *Iris germanica*, and *Mentha viridis* were being used in cosmetics.

In the middle ages, particularly between 16th and 18th centuries, monasteries and churches took over the skills of healing and cultivation of medicinal plants, thereby; preparation of drugs were mostly restricted to a few handful of plants like sage, mint, anise, savory, tansy, and Greek seed (Tucakov, 1990). The silk route trade relations introduced the Arabs to numerous new plants in pharmacotherapy, mostly from India, and they used deadly nightshade, aloe, coffee, henbane, ginger, saffron, strychnos, pepper, curcuma, rheum, cinnamon, senna, and so forth. The European physicians consulted the Arab works, for instance; *De Re Medica* by John Mesue (850 AD), *Canon Medicinae* by Avicenna (980–1037), and *Liber Magnae Collectionis Simplicum Alimentorum Et Medicamentorum* by Ibn Baitar (1197–1248), in which over 1000 medicinal plants were documented (Tucakov, 1965). Though traditional people still used medicinal plants primarily as simple forms; the demand for compound drugs was increasing. During these period, simple pharmaceutical preparations by infusions, decoctions, and macerations using medicinal plants was also introduced and the compound drugs mainly contained of medicinal plants, rare animals, and minerals (Bojadzievski, 1992; Toplak Galle, 2005). Meanwhile, the great expedition by Marco Polo (1254–1324) and by Vasco dè Gama (1498) helped Europe develop rich cultivation of new medicinal plants like Cinchona, Cacao, Ipecacuanha, Ratanhia, Jalapa, Podophylum, Lobelia, Vanilla, Senega, tobacco, Mate, red pepper, as well as quinine bark *Cinchona succirubra*.

Early 19th century was marked by the beginning of scientific pharmacy. The earliest report of chemistry of NP was heralded by the work of Friedrich Wilhelm. With improved knowledge of chemistry and pharmacy; discovery, characterization, and isolation of lead compounds from the medicinal plants were available. Alkaloids were isolated from poppy (1806), ipecacuanha (1817), strychnos (1817), quinine (1820), pomegranate (1878), and other plants. Adam Serturner (1803), a German pharmacist, was known to sequester morphine from opium poppy (*Papaver somniferum*) (Huxtable and Schwarz, 2001). Glycosides, tannins, saponosides, etheric oils, vitamins, hormones, and so on, were also characterized and substantiated (Dervendzi, 1992).

An impediment was observed in the use of medicinal plants as drugs in the late 19th and early 20th centuries, owing to shortcomings incurred during the process of drying of medicinal plants. Moreover, synthetic preparations of alkaloids and glycosides were being used in therapeutics. Much effort was devoted to study the cultivation of medicinal plants which ascertained numerous forgotten plants like *Aconitum*, *Hyosciamus*, *Punica granatum*, *Stramonium*, *Filix mas*, *Secale cornutum*, *Opium*, *Colchicum*, *Styrax*, *Ricinus*, with more stabilization methods being proposed, especially the ones with labile medicinal components. Laws on Drugs and Medical Devices (2007) legislated in the Republic of Macedonia clearly lay rules and regulations for the preparation of herbal drugs, herbal processed products, and traditional herbal drugs using dry and fresh parts of medicinal plants. The current law also has provision for use of herbal substances in homeopathic drugs and can be bestowed without a medical prescription, as over the counter preparations.

1.2 INTRODUCTION

Mankind has been on the pursuit of drugs from nature since time immemorial to alleviate pain and cure illness. Man has always been in harmony with nature to explore and harness NPs as healers of various diseases. Innumerable documentation either written or inscribed has been found to commemorate the usage of NPs as "lead compounds" to improve the quality of life. A lead compound is a novel natural chemical entity that has the potency to be a therapeutic agent by optimizing its pharmacokinetic parameters and minimizing its side effects.

It is of the highest prominence, given the random nature of discovery and the fake unfeasibility of innumerable invention of new active principles, those decision-makers in the pharmaceutical industry should employ definitive strategies and that they must comprehend that these guidelines are not mutually exclusive. Hurriedly established any outcomes may lead to recognition of poor research wherein a brilliant study may remain dormant. Every possible effort should be made in the direction to study the molecular mechanism of action (MoA) of a lead compound once it is discovered and characterized. Therefore, a five-step approach should be embraced for the discovery of new lead compounds as possibly new drugs. These consist of the enhancement of already prevailing drugs, of methodical screening, of retroactive manipulation of biological information, of challenges toward coherent design, and of the use of the target protein structural data. In conclusion, all approaches ensuing in documentation of lead compounds are a priori equally good and prudent provided that the research they persuade is done in a cogent manner (Wermuth et al., 2015). A very recent study also has harnessed to target against cytochrome bc1 and dihydroorotate dehydrogenase by tapping onto the chemical diversity offered by the chemoprotective drugs (Antonova-Koch et al., 2018).

The broad chemistry know-how and industrial sustenance is of utmost importance once a lead compound is recognized. Optimization using medicinal chemistry for the drug development process is the most pertinent factor to develop the best pharmacokinetic profile leading to not only the desired formulation but also the preferred route of administration. Regrettably, this is the step where many possible therapeutic drugs miscarry as it needs to be coupled to biological assays for efficacy, safety, and pharmacokinetics. Therefore, manufacturing processes and procedures should include development of sufficient amount of lead compounds for preclinical evaluation and Phase I/II clinical trials (David and David, 2009).

Cohabituation of host–pathogen throughout evolutionary timeline has led humans to harvest a myriad of NPs from nature (plants, marine organisms, animals or microorganisms) as medical and therapeutic targets. (Butler, 2005; Newman and Cragg, 2007, 2016; Butler, 2008; Cragg et al., 2012). NPs are often stereochemically complex molecules marked with diverse functional groups and high specificity with biological targets. Thus, they are valued as health products or structural templates for drug discovery.

In the current scenario, 36% of all 1073 small-molecules-approved drugs are derivatives of NPs (Newman and Cragg, 2007; Newman and

Gordon, 2012). Data from World Health Organization (World Health Organization, 2009) establish that 80% of the world's population mainly from developing counties still relies mostly on traditional medicines for their primary health care. In advanced countries too, 119 chemical substance used as drugs are obtained from an odd 90 plant species and are a product of isolation of active chemicals from plants used in traditional medicines (Farnsworth et al., 1985). Over 68% of all antiinfectives (antifungal, antibacterial, antiviral, and antiparasitic compounds) are classified as naturally derived (Mukherjee et al., 2001; Lee, 1999), whereas 79.8% of compounds alone are used in cancer treatment (Paterson and Anderson, 2005; Koehn and Carter, 2005; Lee, 2004; Butler, 2004, 2005). Though much focus had been on anticancer drugs, NPs hold great promise in treating vector-borne disease (VBD) and it's the current global need where nearly half of the world's population is infected with at least one type of vector-borne pathogen (Cragg et al., 2011). More than 700,000 annual deaths are accounted globally; wherein more than 17% of all infectious diseases are VBDs (Jones et al., 2008). In over 128 countries, more than 3.9 billion people are at risk of suffering from dengue, with 96 million cases assessed per year. Malaria causes more than 400,000 deaths every year globally, most of them children under 5 years of age. Worldwide, hundreds of millions of people are affected by other diseases such as Chagas disease, leishmaniasis, and schistosomiasis (CIESIN, 2007; WHO, 2004). The risk of being afflicted by vector-borne pathogens has increased their severity with climate and environmental change due to increase in land use (2008).

Billions of people are at risk from vector-borne infectious diseases, such as dengue fever, malaria, yellow fever, and plague transmitted by mosquitoes, ticks, fleas, and other vectors. These diseases intensely restrict development in countries, many of which are located in the tropics and subtropics, with the highest rates of infection rendering poor socioeconomic status. India, being a tropical developing country, faces high fatality in rural as well as urban areas owing to VBD and there is an imperative need for cost-effective, eco-friendly, and safe drugs harnessed from NPs. The most widely known VBD globally is malaria and India still face huge challenge in producing effective drugs against malaria. However in recent times, dengue, which is caused by an arbovirus virus, has become a major public health challenge (Bueno-Marí, 2013).

In this chapter, the authors have restricted in providing a complete account of lead compounds from NPs as potential drugs against malarial parasites. Any account of use of NPs as potential drug targets in other VBDs have been consciously omitted as it was proving to be an over extensive study. Thereby, to endure within the scope of this chapter, the contents are primarily focused on the categorization, their MoA, and possible drug targets from natural lead compounds as Antimalarials only.

1.3 LIFE CYCLE OF MALARIAL PARASITE

Widely spread over the tropical and subtropical regions, including much of Asia, Sub-Saharan Africa, and the Americas, Malaria is a mosquito-borne infectious disease of humans caused by eukaryotic protists of the genus *Plasmodium*. It exhibits a complex life cycle that involves alternating cycles of sexual development (sporogony) in female *Anopheles* mosquitos (definitive host) and asexual division (schizogony) in human (intermediate host). This phenomenon of the malarial parasite to display an alternation of hosts makes it more formidable to understand its life cycle (Fig. 1.1) for effective prevention and eradication of the disease.

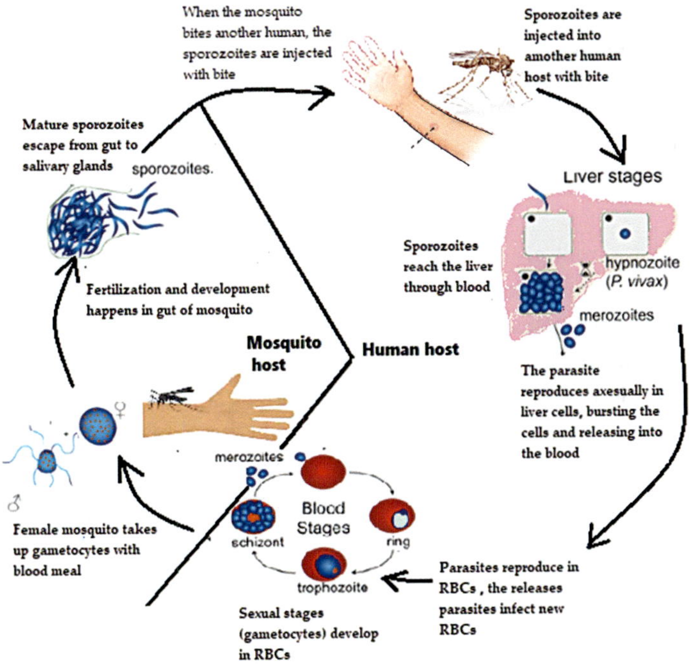

FIGURE 1.1 Sexual and asexual life cycle of *Plasmodium* (adapted from Lee et al., 2014).

1.3.1 HUMAN CYCLE

The salivary glands of female *Anopheles* mosquito host the sporozoites that are the infective form of the parasite. On biting, the proboscis of the mosquito pierces the skin and injects saliva containing sporozoites directly into the bloodstream. The parasite cycle in man includes the following:

1. Primary exoerythrocytic or pre-erythrocytic schizogony
2. Erythrocytic schizogony
3. Gametogony
4. Secondary exoerythrocytic or dormant schizogony

1.3.1.1 PRIMARY EXOERYTHROCYTIC OR PRE-ERYTHROCYTIC SCHIZOGONY

The sporozoites leave the bloodstream within one hour and enter parenchyma cells of liver. Inside the liver cells, the sporozoites, which are elongated and spindle shaped, become rounded successively through multiple nuclear divisions to develop into primary exoerythrocytic schizont. Primary erythrocytic schizogony consists of only one generation. After completion, the liver cells rupture and release merozygotes into the bloodstream.

1.3.1.2 ERYTHROCYTIC SCHIZOGONY

The parasites multiply during the erythrocytic phase and release the merozites which are accountable for carrying out the clinical assault of malaria. The liberated merozygotes attack red blood cells (RBCs) of humans where they multiply and pass through the stages of trophozoites, schizonts, and merozoites. Though the infection tends to die out, erythrocytic schizogony may continue for a considerable period of time. In *P. falciparum* forms of malaria parasites, the erythrocytic schizonts aggregate in the capillaries of the other internal organs as well as in brain so that only young ring forms are found in peripheral blood.

1.3.1.3 GAMETOGONY

Malarial parasites undergo erythrocytic schizogony for a certain period; thereafter, some merozygotes develop into male and female gametocytes known as microgametocytes and macrogametocytes within RBCs of internal organs like spleen and bone marrow. Only mature gametocytes, which are found in the peripheral blood, do not cause any feverish condition in the human host but are produced for the proliferation and perpetuation of the species.

A variety of factors stimulate gametocytogeneis, including hyperparasitaemia, anemia, and antimalarial drug treatment but it appears that the parasite is capable of sensing hostile conditions and transform into gametocytes before preparing to escape into a new host. The microgametocytes of all the four species of *Plasmodium* are smaller in size and are hosted by the carrier, in this case, the man. Although the longevity of mature gametocytes may surpass several weeks, their half-life in the bloodstream may be only 2 or 3 days till these are picked up by the mosquito.

1.3.1.4 SECONDARY EXOERYTHROCYTIC OR DORMANT SCHIZOGONY

Some sporozoites of *Plasmodium vivax* and *Plasmodium ovale* enter into a resting (dormant) stage within the hepatocytes before undergoing asexual multiplication; while others undergo multiplication without delay. This resting stage of the parasite is rounded, 4–6 μm in diameter, uninucleated, and is known as hypnozoites. Hypozoites are reactivated to secondary exoerythrocytic schizonts and release merozoites after remaining dormant for over period of weeks, months, or years (usually up to 2 years). New invasion from liver Merozoites cause infection in the RBCs producing relapse of malaria, a situation in which the erythrocytic infection is eliminated.

1.3.2 MOSQUITO CYCLE

In the stomach of the mosquito, one microgametocyte gives rise to eight thread-like filamentous structures called microgametes by the process of exflagellation. It progresses to macrogamete which fertilizes into a zygote. The zygote elongates and develops into ookinete, a motile vermiculate stage that penetrates the epithelial lining of the mosquito stomach and comes to lie between the external border of the epithelial cell and the peritrophic membrane developing into Oocytes. Oocytes increases in size to reach a diameter of 40–50 m forming sporozoites. The oocyst gets fully mature by the 10th day and ruptures releasing sporozoites within the body cavity of the mosquito. These sporozoites are distributed to all organs of the body via the body fluid but they have a special preference for salivary glands and ultimately reach in maximum numbers in the salivary ducts. At this stage, the mosquito is capable of spreading the infection to man.

Only the sporozoites, merozoites, and ookinetes, which are designed for the invasion of the hepatocytes, erythrocyte, or midgut epithelial cell of the mosquito, respectively, possess a surface coat and the specialized apical end characteristic of apicomplexa. Other stages of the life cycle, which are designed for growth and development within the host cell, lack these invasion organelles.

1.4 ANTIMALARIAL—SIDE EFFECTS, PRECAUTIONS, AND CONTRAINDICATIONS

In recent years, an upsetting rise in the number of malarial deaths is principally due to the ineffectiveness of available drugs that has led to the transmission of multidrug-resistant strains of *Plasmodium*. The disease manifestations due to parasites within red blood cells cause symptoms that typically include fever, headache, and in severe cases progress to coma, and even death. Malaria has traditionally been treated with quinolones such as chloroquine and mefloquine (Fig. 1.2) and with antifolates such as Fansidar (sulfadoxine and pyrimethamine) (Fig. 1.3) (Elujoba, 2005). Unfortunately, most *Plasmodium falciparum* strains have now become resistant to chloroquine and some, such as those in Southeast Asia, have also developed resistance to mefloquine and halofantriene. Moreover, side effects like nausea, vomiting, headache, abdominal pain, vivid dreams, insomnia, and so on are reported with antimalarial. Women should avoid pregnancy while on chloroquine medication as such antimalarial sometimes may lower blood glucose level and may cause dizziness, confusion, increased hunger, nervousness, or anxiety. Quinolone class of drugs is associated with central nervous system toxicity as well adverse psychiatric effects have also been reported (Nevin and Croft, 2016).

1.5 PREVENTION OF DRUG RESISTANCE IN MALARIA

Nowadays, the most important target research areas for public institutions and private companies accountable for mosquito control are the development of natural parasiticides with low toxicity to humans and nontarget organisms of aquatic environments. Since decades, almost exclusive use of chemical insecticides to manage mosquito pests has resulted in mosquito resistance. Resistance may have developed due to mutations in the target site precluding the insecticide–target interaction or changes in enzyme systems of mosquitoes, resulting in more rapid detoxification or sequestration of the insecticide (Hemingway et al., 2004).

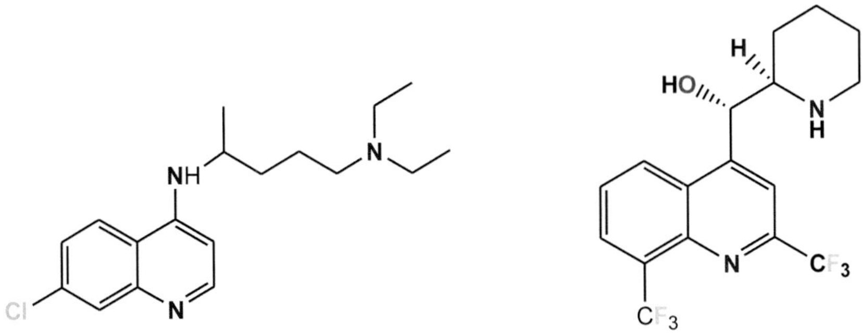

1 **2**

Chloroquine **Mefloquine**

FIGURE 1.2 2D structures of quinolones such as chloroquine and mefloquine.

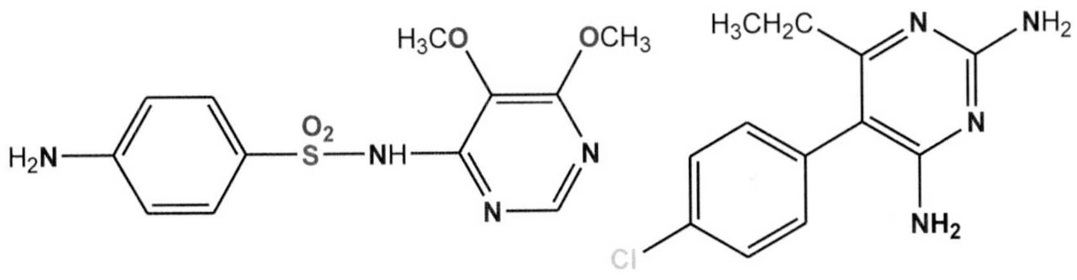

3 **4**

Sulfadoxine **Pyrimethamine**

FIGURE 1.3 Chemical structure of antifolates such as Fansidar (sulfadoxine and pyrimethamine).

As the development of newer drugs appears to take more time than parasitological resistance, the malarial parasites do get ample time to evolve to resist almost all available antimalarials. “Bi-therapy,” which combines artesunate (a derivative of artemisinin, recommended by the World Health Organization) with Malarone (or Malanil) has been shown to be very useful in slowing down development of resistance. Artemisinin derivatives used as antimalarial combination therapies have become part of first-line references as the optimum approach to malaria chemotherapy, mainly in countries in which multidrug-resistant malaria has been widespread (White et al., 1999; Trape et al., 2002). Till date, the MoA of artemisinin was thought to damage the parasitic proteins by generating free radicals activated by heme; but few new recent theories have identified many other MoAs for artemisinin action. Studies have shown that not only heme, but PfATP6 (Ca^{2+} transporter) as well phosphatidylinositol-3-kinase as potential MoAs. Artemisinin is a potent inhibitor of *P. falciparum* (Shandilya et al., 2013; Mok et al., 2015) and showed decreased parasite development due to up-regulation of the unfolded protein response pathways (Mbengue et al., 2015). The exploration of these novel MoAs will pave way for the development of future antimalarials using In-vitro evolution and whole-genome analysis. These are based on genome-based target discovery methods on compounds identified via phenotypic screens (Tse et al., 2019).

Sadly, even as the numbers of active malarial drugs are inadequate; the selection of resistant strains is further narrowing the number of effective molecules. Moreover, the failure of the artemisinin-based combinations of drugs mainly in South-Asian counties have compelled the scientists to reevaluate the current approaches to combination therapies for malaria with incorporation of three or more drugs in a single treatment (Ashley and Phyo, 2018). Although antimalarial hybrids have low toxicity and better pharmacokinetics, the therapeutic choices are still too limited (Agarwal et al., 2017). As safe alternatives to chemical or bacterial insecticides, major emphasis must be applied on the use of natural plant-based products as larvicides to try to minimize resistance problems (Sundaravadivelan et al., 2013). As of now, more than 2000 plant species are known to produce chemical factors and metabolites with pesticidal value and products of some 344 species have been reported to act against mosquitoes (Sukumar et al., 1991). Even though very few plant products act against mosquitoes as growth regulators, repellents, and ovipositional deterrent, none are developed to control mosquitoe breeding (Amer and Mehlhorn, 2006; Rajkumar and Jebanesan, 2007). Nevertheless, few plant extracts show antiviral activity against Dengue (Tang and Ooi, 2012), Yellow Fever (Ojo et al., 2009), and Ross River Fever (Semple et al., 1998) which are also mosquito-borne diseases.

The reliable source of accessibility changes that are due to seasonal and environmental factors as well as the threat for extinction of certain NPs are among the few challenges that are faced by NPs antimalarials. Also the complexity of mixtures after fractionation, isolation of small quantities of bioactive substances and the chemical challenges like stability and solubility poses other hindrances. But still new pursuits are being made to generate antimalarials drugs from natural sources with structurally diverse compounds and a variety of chemical classes (Guantai and Chibale, 2011) (Fig. 1.4).

1.6 LEAD COMPOUNDS AS ANTIMALARIALS

Quinine from *Cinchona* sp. has been adjusted as one of the popular naturally occurring lead compounds that are derived from plants. These widely used antimalarials, namely, choloroquine and sulfadoxine-pyridoxine, are losing their effectiveness owing to resistant varieties of malarial parasites. Thus, there is a pressing need to ascertain other natural lead compounds to ensure their effectiveness against resistant *P. falciparum* strains. Also, these compounds should ensure good compliance, should remain effective for a considerable time period, affordable, safe, and with noninvasive route of administration. Apart from artemisinin from plant sources, marine-derived lead compounds are the new breakthrough antimicrobilas. Marine organisms comprising of invertebrates like sponges, tunicates, soft corals have incredible potential to produce abundant variety of secondary metabolites as well as some unique features of the marine environment. Based on their chemical structures, NPs lead compound antimalarials have been divided into different classes: endoperoxides, isonitrile-containing derivatives, alkaloids, nonalkaloids.

1.6.1 MALARIAL LIFE CYCLE AFFECTED BY NATURAL LEAD COMPOUNDS

Malaria is till date considered as a deadly disease because popular antimalarials usually target the Schizont and ring-form trophozoite of the disease-causing *Plasmodium* parasites. One must apprehend that for

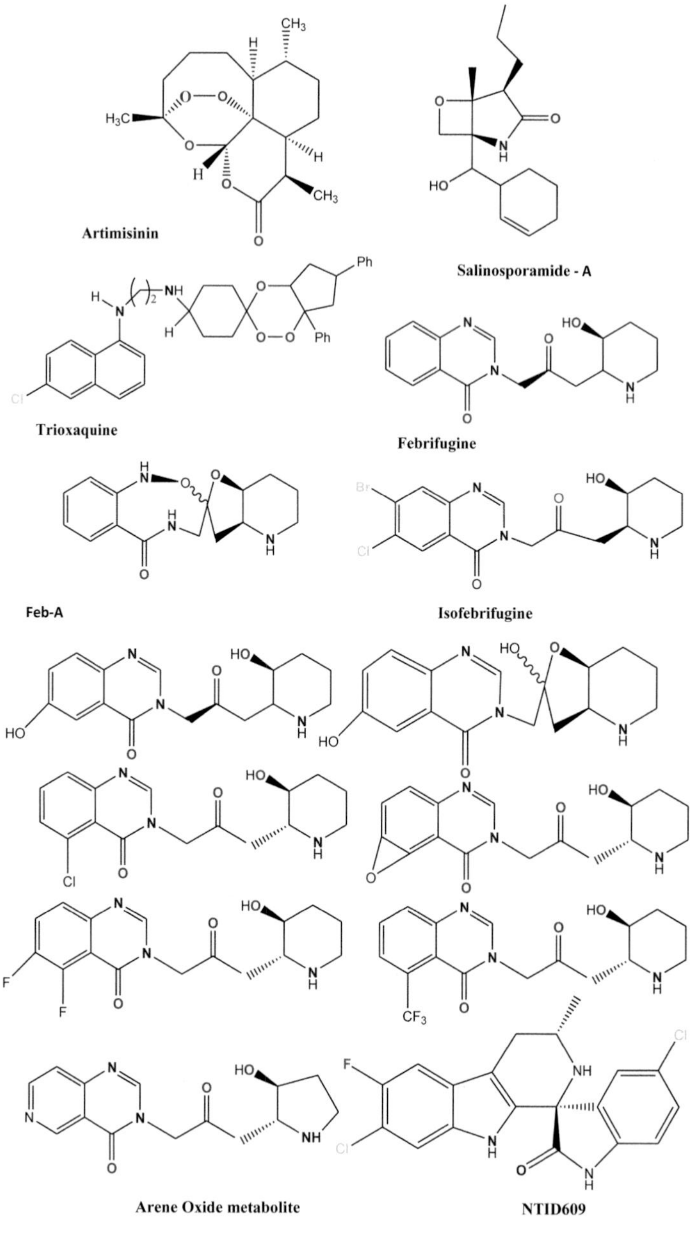

5

FIGURE 1.4 2D structure of diverse compounds from different chemical classes.

complete annihilation of malaria, drugs that target and block transmission of the parasite from mosquito to humans and vice versa should be considered; that means gametocytes and the sporozoites stage should be targeted. Research should now be based on drugs with transmission-blocking capabilities rather than targeting a specific stage.

Among the NPs lead compound antimalarials, only endoperoxides, namely, OZ439 showed promise in inhibiting gametocyte maturation and gamete formation. Further, some permitted drugs, such as pyronaridine and atovaquone, target both asexual blood stages as well as sexual stages of *Plasmodium*. Endoperoxidases (artemisinin, DHA, artesunate, OZ277, and OZ439) also target asexual stages while some others like 8-aminoquinoline and NPC-1161B are reported to inhibit sporogony. One study showed that some of endoperoxidases showed lethality for mature gametocyte, liver schizont, and sporogonic parasite stages that lack hemoglobin metabolism (Fig. 1.5). Thus, endoperoxide has lesser impact on ookinete and oocyst development but have profound activity on exflagellation, oocyst production, and liver schizont development (Delves et al., 2012).

1.6.2 ENDOPEROXIDES

Endoperoxide compounds (natural/semisynthetic/synthetic) possess varied structural conformation and that makes them very desirable as antimalarial agents against resistant *P. falciparum* malaria. The pharmacodynamic potential of endoperoxide-based antimalarials is due to lack of the lactone ring in the 1,2,4-trioxane ring system which changes the structural scaffolding of endoperoxide. Thus, in recent years, antimalarial drugs were developed against resistant malaria by using 1,2,4-trioxane-, 1,2,4-trioxolane-, and 1,2,4,5-teraoxane-based scaffolds. This diversity of endoperoxide molecules, including their chimeric (hybrid) molecules, provides potent antimalarial agents.

1.6.2.1 ENDOPEROXIDASES AS ANTIMALARIALS FROM PLANT SOURCES

1.6.2.1.1 Artemisinin

Chinese folk medicine used a herbal remedy isolated from *Artemisia annua* (Elujoba, 2005) is basically an endoperoxide sesquiterpene possessing a

1,2,4-trioxane moiety, named artemisinin (Qinghaosu) (Rustaiyan et al., 2016). This endoperoxide linkage is an essential feature for antimalarial activity of artemisinin and of its derivatives (e.g., the oil-soluble artemether and the water-soluble artesunate) (Avery et al., 1992). Organic chemists harness certain unique structural features like peracetal, acetal, and lactone functional groups of artemisinin to develop antimalarials. Artemisinin and several semisynthetic derivatives (Fig. 1.6) meet the dual challenges posed by drug-resistant parasites and rapid progression of malarial illness (Hien and White, 1993).

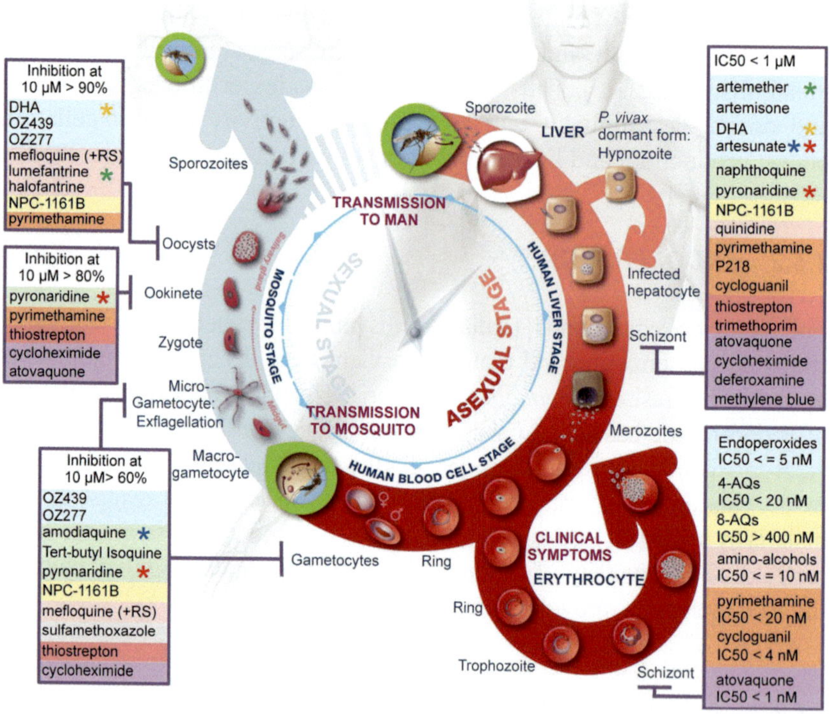

FIGURE 1.5 Effect of antimalarials' mainly on liver stage, blood stage, and vector stage of the life cycle of *Plasmodium*. Encircled in Green, the key entry points leading to transmission of the parasites from vector to host and from host to vector are indicated.

Source: Reprinted from Delves et al. (2012). Creative Commons Attribution License.

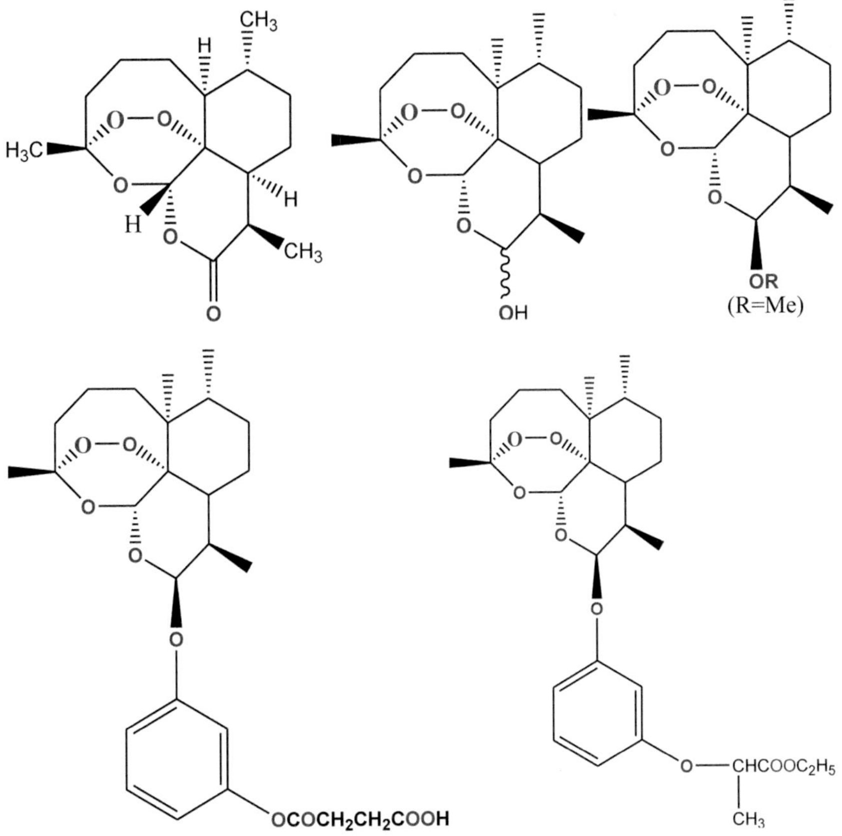

6

FIGURE 1.6 Chemical structures of artemisinin and its derivatives.

1.6.2.2 ENDOPEROXIDASES AS ANTIMALARIALS FROM MARINE ORGANISMS

Endoperoxide derivatives from marine sources are now been seen as valuable alternative to artemisinin. According to their biogenetic origin, these molecules have been allocated in two categories: polyketides and terpenoids.

1.6.2.2.1 Polyketide

Polyketides are marine sponges (family Plakinidae) which possess simple endoperoxide with six- or five-membered 1,2-dioxygenated rings (1,2-dioxane or 1,2-dioxolane, respectively). Some of them also contain 3-methoxy substitution, building a peroxyketal group. The plakortin analogs, namely, dihydroplakortin, 3-epiplakortin, plakortide Q (Fig. 1.7), show good antimalarial activity against D10 (chloroquine-sensitive strain) and W2 (chloroquine-resistant strain) of *P. falciparum*. The simple chemical structure poses a good probe to study mechanisms of action as well as structure–activity relationships. The role of the conformational behavior and the western alkyl side chain of the dioxane ring in interacting with heme planar target is the criterion on which a series of semisynthetic derivatives of plakortin has been prepared (Fattorusso et al., 2006).

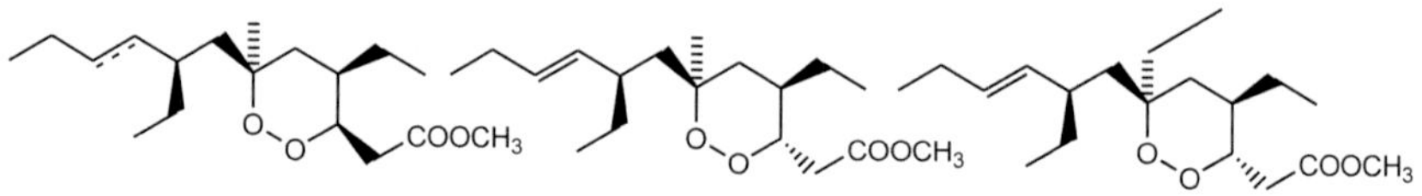

7 **Dihydroplakortin** **8** **3-Epiplakortin** **9** **Plakortide Q**

FIGURE 1.7 Chemical structures of plakortin: dihydroplakortin, 3-epiplakortin, and plakortide Q.

1.6.2.2.2 Terpenoids

1,2-Dioxane derivatives, also known as endoperoxide-containing terpenoids isolated from marine sources have also been verified for their antimalarial activity. Sigmosceptrellin A, a norsesterterpene derivative, is presented to be active against *P. falciparum* (IC50 ~ 450 ng/mL) but the C-3 epimer of the same molecule, named sigmosceptrellin B, proved to be four times less potent (Fig. 1.8). Thus, this tells us the significance of stereochemistry to ascertain the antimalarials activity (D'Ambrosio et al., 1998).

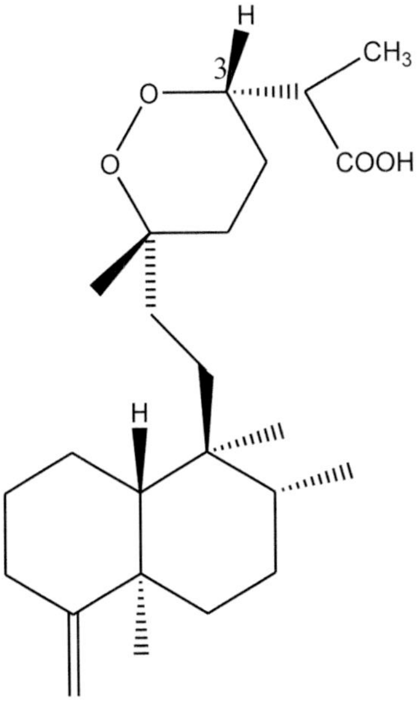

10

Sigmosceptrellin A

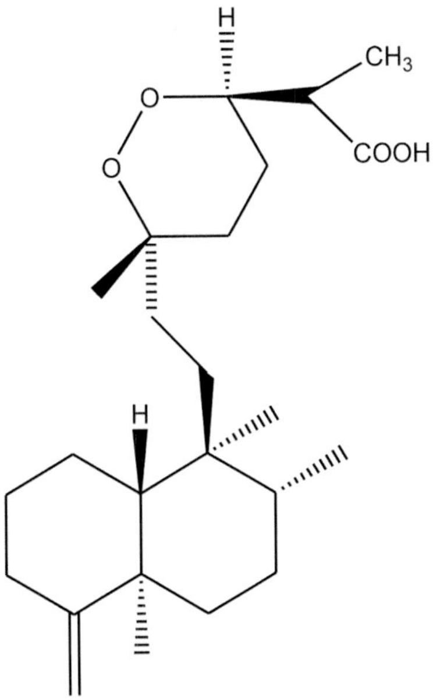

11

C-3 Epimer Sigmosceptrellin

FIGURE 1.8 Chemical structures of sigmosceptrellin A and of its C-3 epimer sigmosceptrellin.

1.6.2.3 MoA OF ENDOPEROXIDES

During the blood-stage phase of *Plasmodium* parasite, more than 70% of the hemoglobin within the infected erythrocyte is digested and heme is released. The antimalarial action of these drugs is based on their potential to target either enzymes (hemepolymerase and Hb-degrading proteases) or non-enzymes (heme/ferriprotoporphyrin). Hematin, upon reduction to heme, gives away ferrous iron [Fe(II)], which in turn reduces the endoperoxide bridge of ART to cytotoxic carbon radical species which are toxic to the parasite through alkylation of sensitive macromolecular targets (Fig. 1.9), thus killing the *Plasmodium* parasites (Kamchonwongpaisan and Meshnick, 1996). A Ca^{2+}-dependent ATPase specific of *P. falciparum* (PfATP 6) has also been suggested as a potential target for all active species (Eckstein-Ludwig et al., 2003). Nevertheless, there are other reasons for weak actions of antimalarials. Most of them are bases which tend to accumulate in the acidic food vacuoles of the parasites. Therefore, to achieve desired effects at molecular and cellular level; not only the release of reactive heme upon Hb-digestion should be the biochemical basis but also basicity (p*K*a) of drug molecules and cytoplasmic/vacuolar pH are other imperative concerns that should be taken into account for coherent drug design of newer antimalarial drugs.

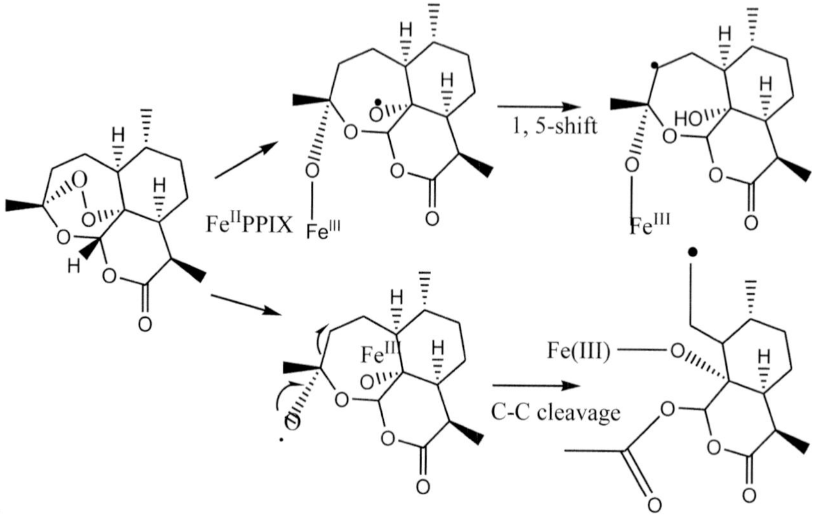

12

FIGURE 1.9 Mechanism of action of artemisinin.

1.6.3 AXISONITRILE

1.6.3.1 ISONITRILE AS ANTIMALARIAL FROM MARINE ORGANISMS

In the early 1970s, marine sponge (*Axinella cannabina*) was the main source to isolate isonitrile secondary metabolites, mainly axisonitrile-1 (**13**); axisothiocyanate-1 (**14**); other isonitrile-, isothiocyanate-, and formamide-containing sesquiterpenoids like axamide-1 (**15**); axisonitrile-2 (**16**); axisothiocyanate-2 (**17**); axamide-2 (**18**); axisonitrile-3 (**19**); axisothiocyanate-3 (**20**); and axamide-3 (**21**) (Fig. 1.10) (Di Blasio et al., 1976).

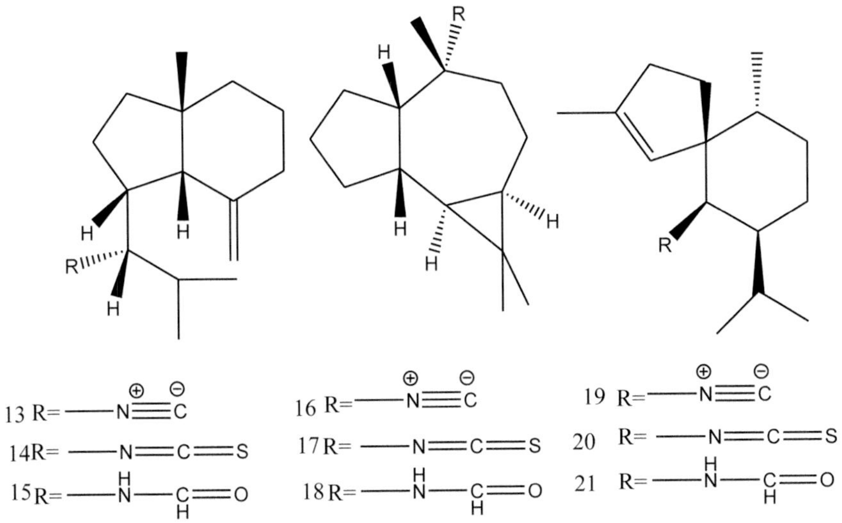

FIGURE 1.10 Isonitrile-, isothiocyanate-, and formamide-containing sesquiterpenoids from the sponge *Axinella cannabina*.

In 1992, axisonitrile-3 (**19**) isolated from the sponge *Acanthella klethra* was shown to be a potent antimalarial on chloroquine-sensitive (D6, 142 ng/mL) as well chloroquine-resistant (W2, 17 ng/mL) *P. falciparum* strains, whereas closely related axisothiocyanate-3 (**20**) was inactive suggesting that isonitrile functional group should be present in carbon backbone (Angerhofer et al., 1992). Marine sponges of the families Axinellidae and Halicondridae came up with a series of diterpenes based on amphilectane, isocycloamphilectane, and neoamphilectane skeletons having isonitrile, isothiocyanate, and the rare isocyanate group (Fig. 1.11). Antimalarials with Isonitrile was also derived from the Japanese sponge *Acanthella* sp. (Fig. 1.12) (Wright and Lang-Unnasch, 2009). Isonitrile kalihinanes showed good potency against *Plasmodium* parasites even at very low nanogram range (IC50 = 0.4 ng/mL). All the derivatives are however functionally more potent depending upon the presence of isonitrile group on the carbon skeleton (White et al., 2004).

1.6.3.2 MoA OF ISONITRILE DERIVATIVES

Axisonitrile-3 and diisocyanoadociane bind to the heme iron (FP) forming a coordination complex, thus preventing destruction of FP through both the peroxidative and glutathione-mediated sequestration of FP into β-hematin. 3D-QSAR with molecular receptor modeling studies has proven that active isonitrile compounds, like the quinoline antimalarials, exercise their anti-plasmodial activity by preventing FP detoxification (Wright et al., 2001).

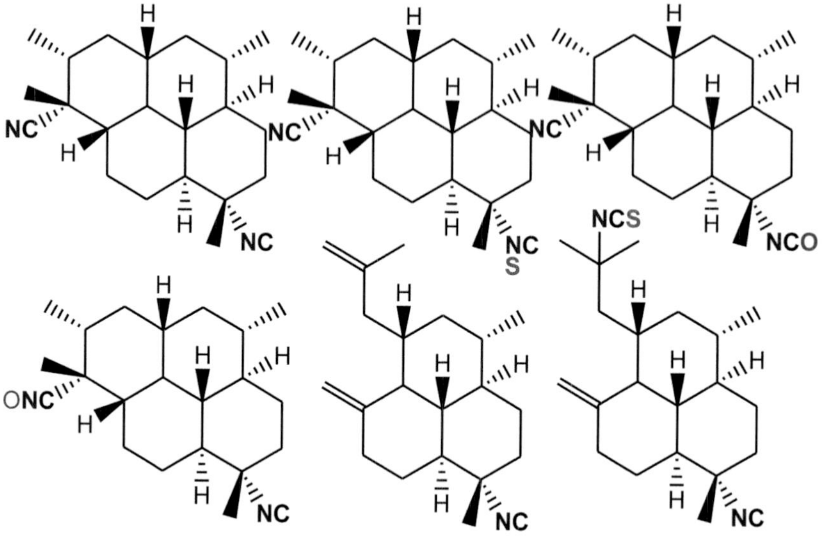

22

FIGURE 1.11 Representative isonitrile- and isothiocyanate-containing diterpenoids isolated from the marine sponge *Cymbastela hooperi*.

1.6.4 ALKALOIDS

Plant defenses are due to their ability to produce large groups of secondary metabolites mainly alkaloids as these molecules are efficient against pathogens and predators due to their toxicity. Alkaloids such as caffeine and nicotine are potent neuroactive molecules, emetine fight oral intoxication, while vincristine and vinblastine are potent antimalarials. However, alkaloids may pose both harmful and beneficial depending on the ecological or pharmacological context depending on specific dosage, exposure time, and individual characteristics, such as sensitivity, site of action, and developmental stage. Therefore, to discover new bioactive molecules within alkaloids so as to use them against targets of interest such as cancer cells, pathogens, herbivores, or unwanted physiological conditions; a complete knowledge of alkaloid biosynthesis and mechanisms of action is crucial.

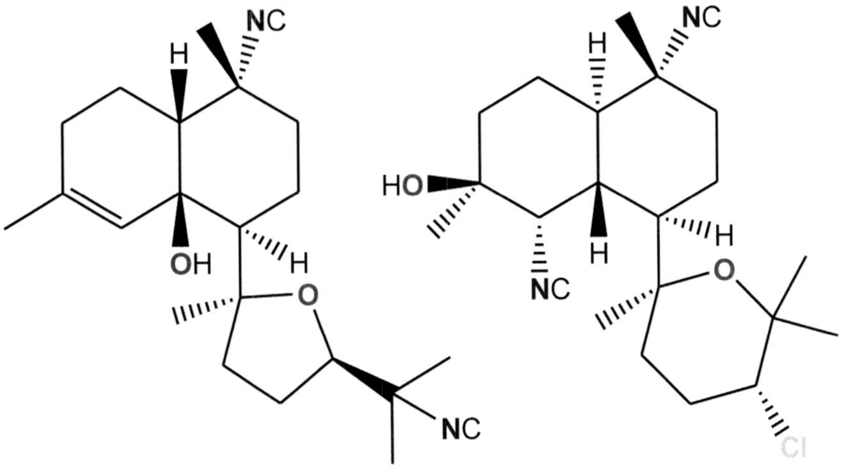

23

FIGURE 1.12 Representative kalihinane diterpenoids isolated from the marine sponge *Acanthella* sp.

1.6.4.1 ALKALOIDS AS ANTIMALARIALS FROM PLANT SOURCES

Quinine, an alkaloid isolated from Cinchona bark in 1820 was indeed the first active component isolated from a natural source to cure malaria. Until the 20th century, many more were developed based on the active structure of Quinine. Later, more than 100 alkaloids from other plants were demonstrated with significant antimalarial activity, some of these even more potent than chloroquine (Saxena et al., 2003). Alkaloids from plants such as *Cyclea barbata*, *Cyclea atjehensis*, *Stephania pierrei*, *Stephania erecta*, *Pachygone dasycarpa*, *Curarea candicans*, *Albertisia papuana* (Menispermaceae), *Hernandia peltata* (Hernandiaceae), and *Berberis valdiviana* (Berberidaceae) showed a wide range of antiplasmodial assays.

1.6.4.1.1 Bisbenzylisoquinolines

Many folklore medicinal plant species, particularly the members of the Menispermaceae, Berberidaceae, Ranunculaceae, Annonaceae, and Monimiaceae, contain a varied group of alkaloids called Bisbenzylisoquinol (Fig. 1.13). The traditional antimalarial plant *Strychnopsis thouarssi* was used to derive a morphinan alkaloid, a biogenetically derivative of benzylisoquinolines. This is a unique class of antimalarial drugs with exceptional inhibitory activity against *Plasmodium* hepatic stages (Carraz et al., 2006). Similarly, another phenolic aporphine-benzylisoquinoline alkaloid derived from the roots of *Thalictrum faberi* Ulbr. (Ramunculaceae) showed antiplasmodial activity (Lin et al., 1999). These alkaloids have only one diaryl ether bridge and show low KB-cell cytotoxicity with high antiplasmodial activity (Fig. 1.13). This activity is predisposed by the change in conformation at chiral centers and by substitution of other functional groups on the aromatic rings. However, loss of both toxicity and antiplasmodial activity was observed with decrease in lipophilicity, as seen in quaternarized or N-oxide derivatives. This loss was maybe a result of changed membrane permeability (Angerhofer et al., 1999).

1.6.4.1.2 Naphtylisoquinolines

Tropical lianas, also well-known traditional plants and belonging to the families Dioncophyllaceae and Ancistrodaceae, comprise more than 70 natural and 150 derivatives of structurally exclusive biogenetically derived acetate alkaloids (Fig. 1.14) (Bringmann and Feineis, 2000). *Plasmodium berghei*-infected mice when administered with dioncophylline C, dioncophylline B, and dioncopeltine A showed complete clearance of parasites without any toxic effects (Bringmann and Rummey, 2003; François et al., 1997). Structurally similar to dioncopeltine A, a new alkaloid called Habropetaline A, from *Tryphyophyllum peltatum* (Ancistrocladaceae) also exhibited strong antiplasmodial activity against *P. falciparum* (Bringmann and Rummey, 2003). Similarly, dioncophylline E, a semistable compound isolated from *Diconcophyllum thollonii* displayed good antiplasmodial activity against *P. falcifarum* parasites. Antimalarials against CQR and CQS strains of *P. falciparum* are effective by korupensamine A, from *Ancistrocladus korupensis* (Bringmann et al., 2001). Certain dimeric natural derivatives, like michellamine B, are also the subject of research as antimalarial (Fig. 1.12) (Bringmann et al., 2001).

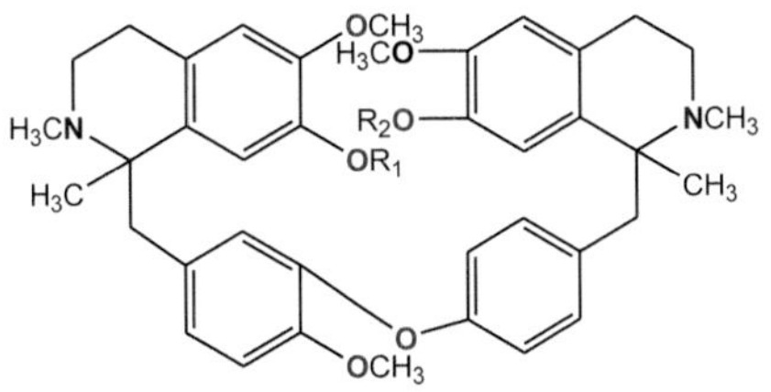

1S, 1'S; R_1= H; R_2= CH_3(+) -neothalibrine
1R, 1'S;R_1= CH_3 = H (+) -temuconine

24

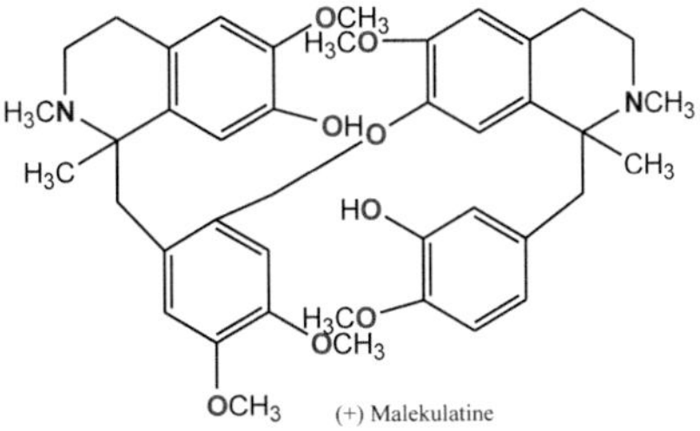

(+) Malekulatine

25

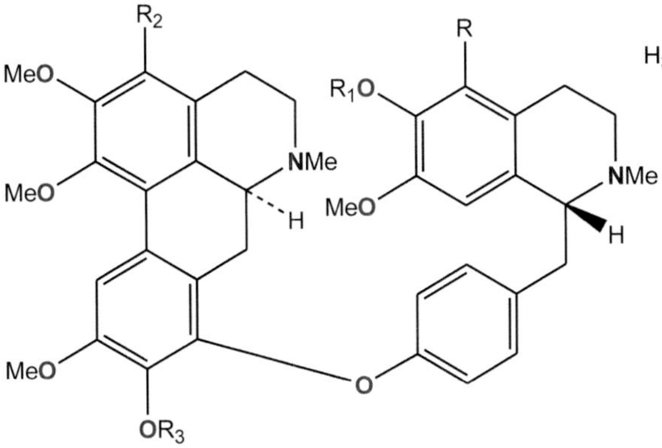

1: R = R_3 = H, R_2= OH, R_3= Me
2: R = R_3= H, R_1= Me, R_2= OH
3: R= R_1= R_2= R_3= H

26

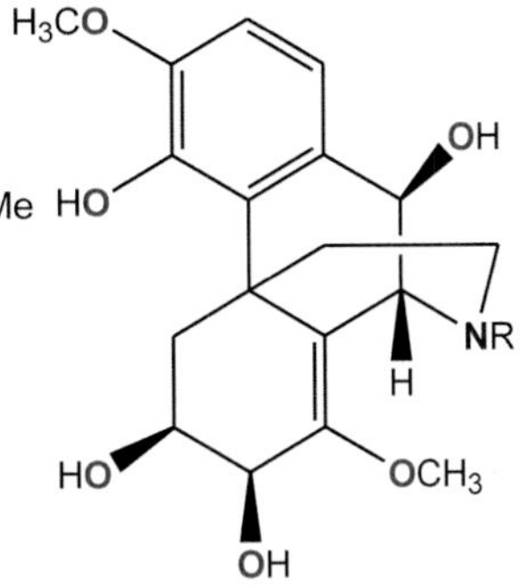

R = H Tazopsine
R = cyclopentyl (semi-synthetic derivative)

27

FIGURE 1.13 Structures of biogenetically derived benzylisoquinoline alkaloids with selective antiplasmodial activity.

1.6.4.1.3 Mono- and Bis-Indole Alkaloids

Indole alkaloids are those NPs used to treat malaria are also obtained from plants mainly of the genera *Strychnos* (Loganiacae) and *Alstonia* (Apocynaceae) (Frederich et al., 2008). These are the alkaloids with a molecular weight higher than 400 and an important steric crowding. They usually are indole analogs of emetine (usambarensine, ochrolifuanine,

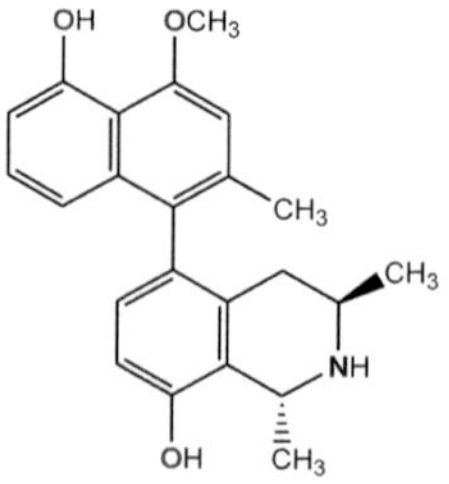

28

Dioncophylline C

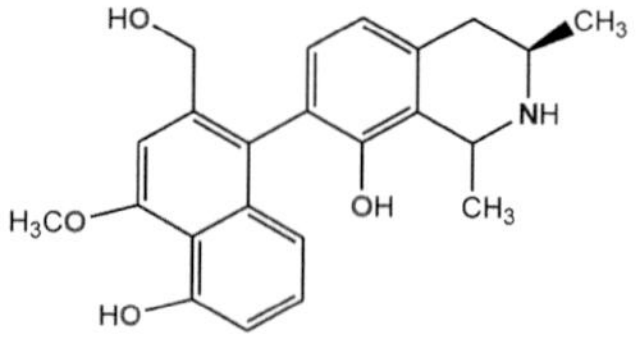

29

Dioncopeltine A

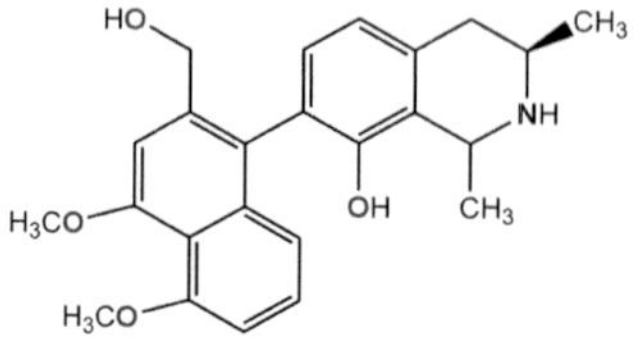

30

Hebropetaline A

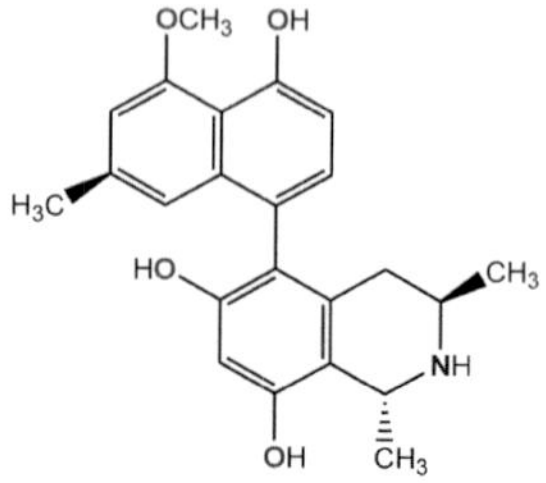

31

Korupensamine A

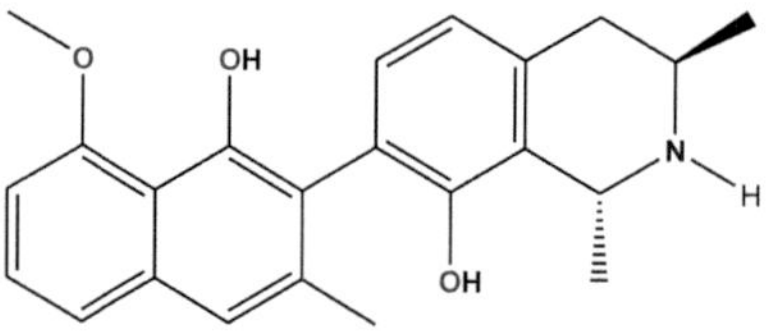

32

Dioncophylline E

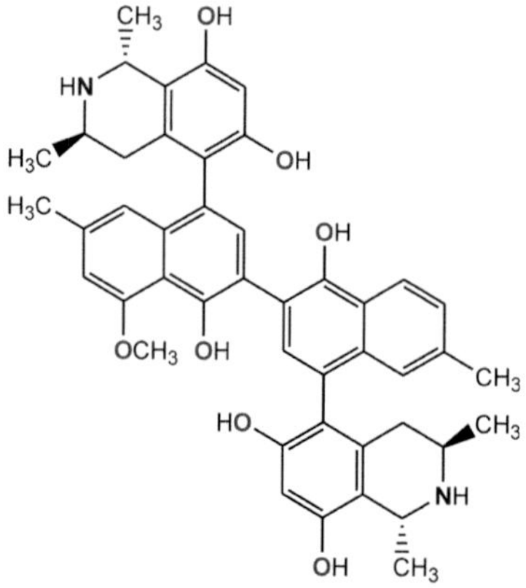

33

Michellamine A

FIGURE 1.14 Structure of naphtyisoquinoline alkaloids: antimalarial (monomeric) and naphtylisoquinoline (dimeric).

strychnopentamine); or bis-indole alkaloids such as voacamine and ergoline derivatives, matopensines and isosungucines (Fig. 1.15). Among these, the most interesting compound seems to be malagashanine (Fig. 1.16) found in *Strychnos myrtoides*. These classes of alkaloids are best effective against

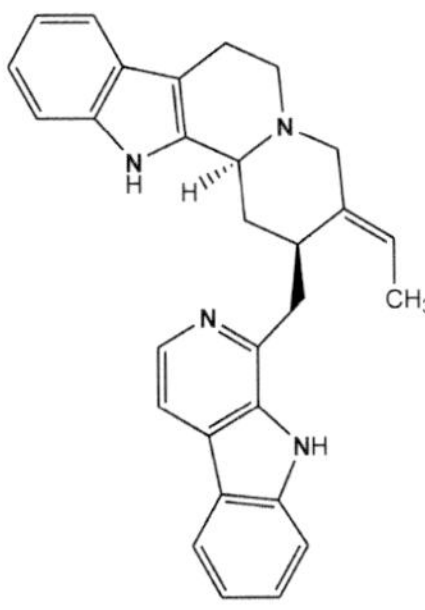

34

Usambarensine

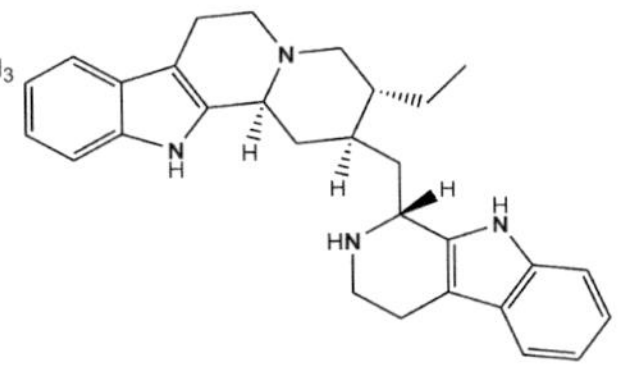

35

Ochrolifuanine

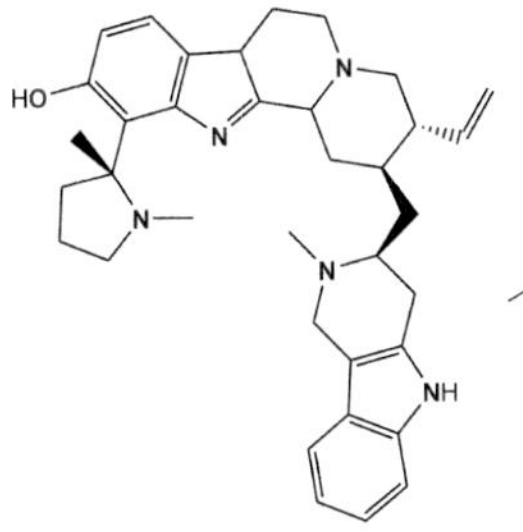

36

Strychnopentamine

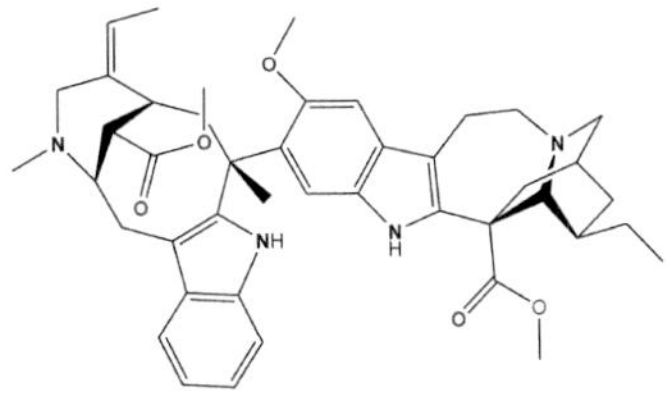

37

Voacamine

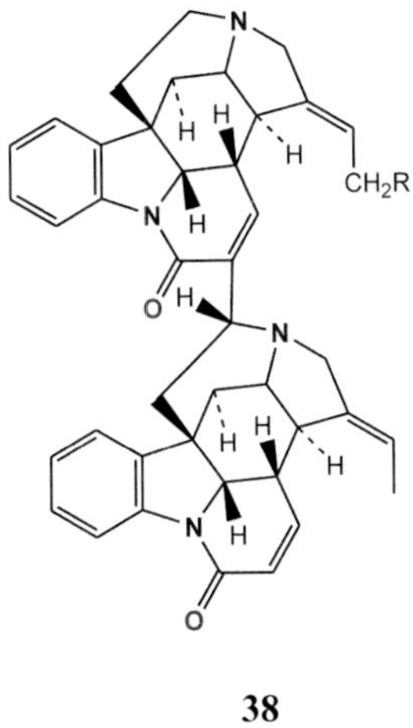

38

Sungucine

FIGURE 1.15 Structure of Indole "analogues" of emetine (*Usambarensine, Ochrolifuanine, Strychnopentamine*) and bis-indole alkaloids (*Voacamine* and *Sungucine*).

chloroquine-resistant strains and some monoindole alkaloids are also able to reverse the resistance to chloroquine.

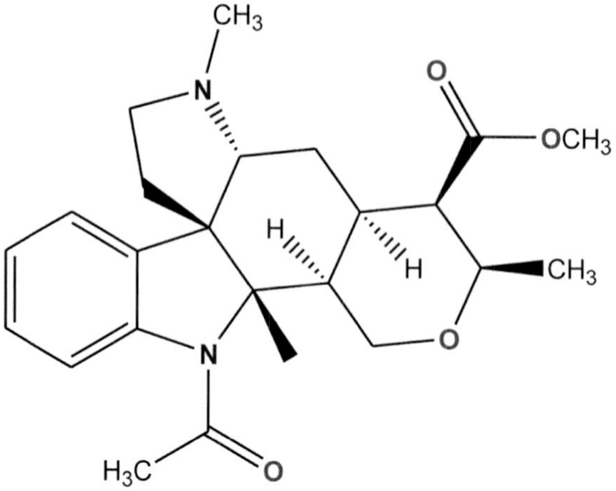

39

Melagashanine

FIGURE 1.16 Structure of monoindole alkaloids: melagashanine.

1.6.4.1.4 Cryptolepine

Roots of *Cryptolepis sanguinolenta* contain cryptolepine, an indoloquinoline alkaloid which was shown to have potent effects against both CQS and CQR *P. falciparum* strains. Cryptolepines are antiplasmodial indole alkaloids containing unsaturated monomeric heterocycles and are chemically much more attainable (Fig. 1.17). Antimalarial activity is also demonstrated by Neocryptolepine, a minor alkaloid, isolated together with cryptolepine. Cryptolepine analogs with addition of nitro group and halogens were also synthesized. One such derivative named 2,7-dibromocryptolepine showed nine-fold greater potency than cryptolepine against CQR *P. falciparum* alone (Jonckers et al., 2002). These molecules exhibit good antimalarial activity by inhibiting the formation of hematin (Wright et al., 2001, Wright, 2005), and intercalating with DNA and stabilizing the topoisomerase II–DNA covalent complex (Cimanga et al., 1997; Wright et al., 2001), thus becoming a favorable lead compound.

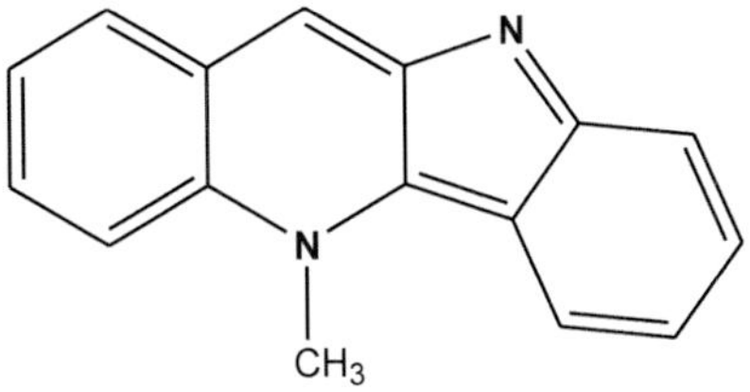

40

Cryptoplne (I)

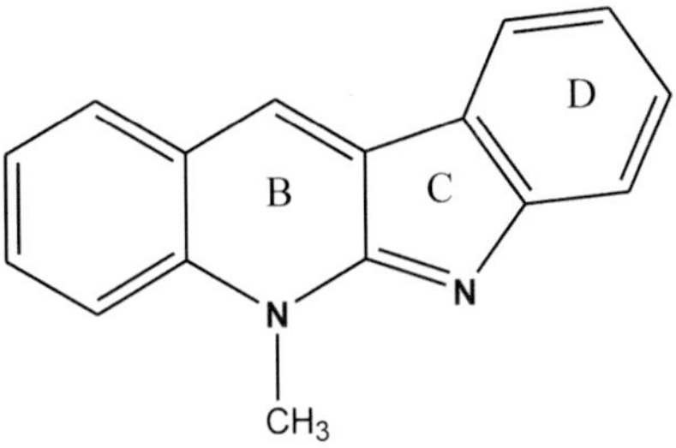

41

Neocryptalepine (II)

FIGURE 1.17 Structure of antimalarial cryptolepine and neocryptolepine.

1.6.4.1.5 *Other Alkaloids*

Alkaloids like benzophenantridines, quinolines, furoquinolines, 2-alkyl-quinolines, and acridines are derived from plants belonging to the Rutaceae family (Fig. 1.18) (Michael, 2003; Waterman, 1999). Benzofenantridine alkaloids, isolated from trunk bark of *Zanthoxylum rhoifolium*, composed of seven fractions, of which nitidine showed potent antiplasmodial activity against *P. falciparum*. Normelicopicine and arborinine alkaloids isolated from *Teclea trichocarpa* (synon. *Toddalia trichocarpa*) though is active against *P. berghei*-infected mice but displayed restricted in-vitro toxicity against *P. falciparum* strains (HB3 and K1). Normelicopicine (Muriithi et al., 2002), 23 furoquinoline, and acridone alkaloids were shown to have effects against CQR (W2) and CQS (HB3) clones of *P. falciparum* (Basco et al., 1994). Tetrahydroquinoline, an alkaloid isolated from the trunk bark of *Galipea officinalis*, is used as tonic and stimulant against fever (Jacquemond-Collet et al., 2002).

1.6.4.2 *ALKALOIDS AS ANTIMALARIALS FROM MARINE ORGANISMS*

1.6.4.2.1 *Manzamines*

Marine taxonomically unrelated sponges belonging to different genera (e.g., *Xestospongia*, *Ircinia*, and *Amphimedon*) and different orders are the source of very complex polycyclic (7–8 rings or more) alkaloids which are reported to have antimalarial activity (Ang et al., 2000). It is hypothesized

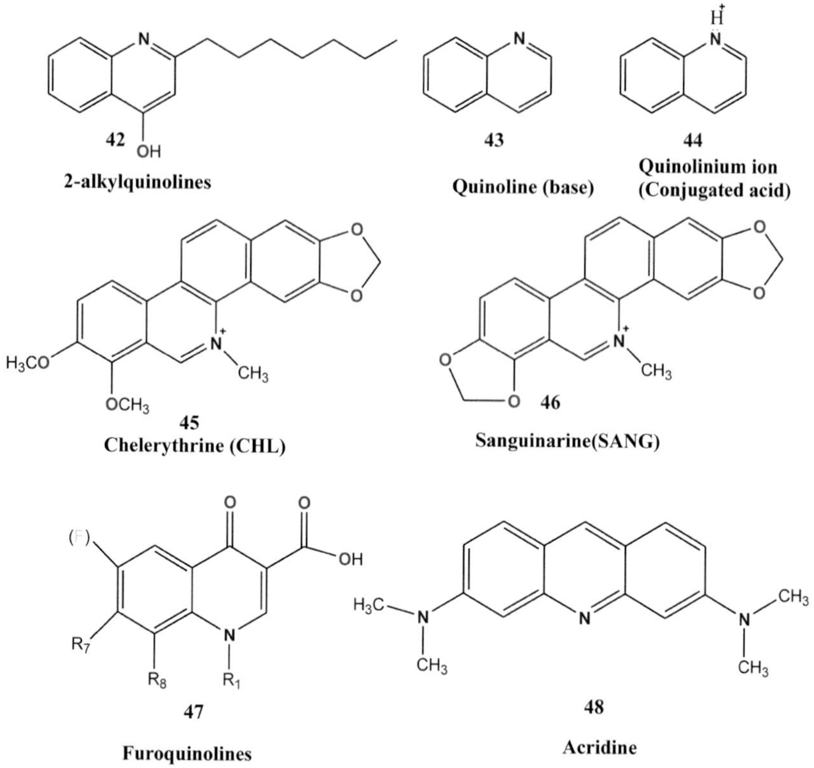

FIGURE 1.18 Structure of different classes of alkaloids from plants belonging to the Rutaceae family (benzophenantridines, quinolines, furoquinolines, 2-alkylquinolines, and acridines).

that manzamines have a symbiotic bacterial origin rather than as true sponge metabolites. More recently, *Micronosphora* sp. was identified as the bacteria producing manzamines (Wypych et al., 2008). Interestingly, Manzamines have differential antimalarial activity. Manzamine A, 8-hydoxy derivative inhibit the growth of *P. falciparum* both *in vitro* (IC50 ~ 5.0 ng/mL) and *in vivo*, whereas manzamine F lacks the potency. It has been noted that any structural changes like insertion of a ketone group or reduction of the double bond, attachment of the hydroxyl group at position 6 instead of 8 all have deleterious effect of antimalarial activity as indicated by the lower potency of manzamine Y (6-hydroxy-manzamine A), exhibiting IC50 ~ 600 ng/mL (Rao et al., 2006). On the other hand, neokauluamine (Fig. 1.19) a very complex molecule constituted by two manzamine units dimerized through ether linkages between the eight-membered rings demonstrated the same antimalarial activity as manzamine A. (El Sayed et al., 2001).

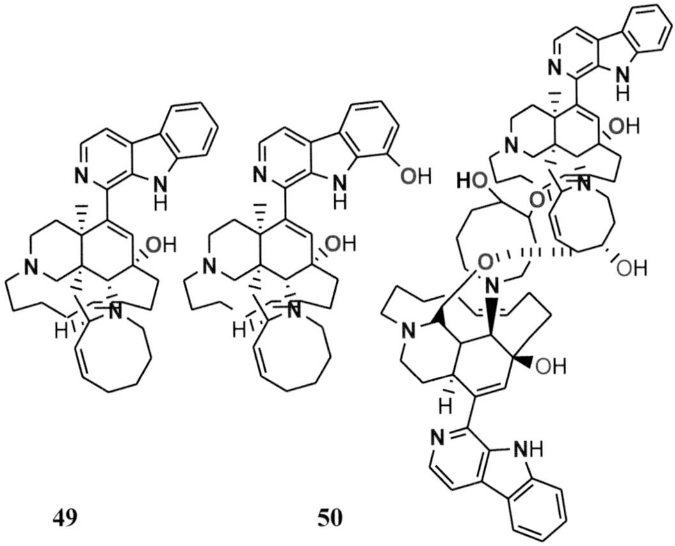

49 **50** **51**

Manzamine-A **Manzamine-F** **Neokauluamine**

FIGURE 1.19 Chemical structure of manzamine A, manzamine F, and *Neo*kauluamine.

1.6.4.2.2 Lepadins

Australian marine invertebrates, *Clavelina lepadiformis* and *Didemnum* sp., are the source of linear eight-carbon chain decahydroquinoline derivatives. Lepadin E displayed a momentous antimalarial activity (IC50 = 400 ng/mL) while its close analogs lepadin B and D are almost completely inactive (Fig. 1.20). The change in activity is due to the placement of 2E-octenoic acid ester in place of the secondary alcohol. This change in conformation may serve to stabilize nonbonding interactions with heme, or with any other receptor molecule (Li et al., 2008).

1.6.4.2.3 Phloeodictynes

These are 6-hydroxy-1,2,3,4-tetrahydropyrrolo[1,2-*a*] pyrimidinium alkaloids derivatives isolated from the haplosclerid sponge Oceanapia

[= Phloeodictyon] fistulosa (Adeline and Debitus, 1992). Similar to lepadins, these are C-6 compounds with an OH group, a variable length alkyl chain and at N-1 a four/five-methylene chain ending in a guanidine group (Fig. 1.21). At least 25 different components, of which 17 are new members with variable terminus and length chains, were characterized, besides less abundant analogs bearing a thioethylguanidine side chain. They are noted to prove active against chloroquine-resistant *P. falciparum* with IC50 values ranging from 0.6 to 6 μM (Mancini et al., 2004).

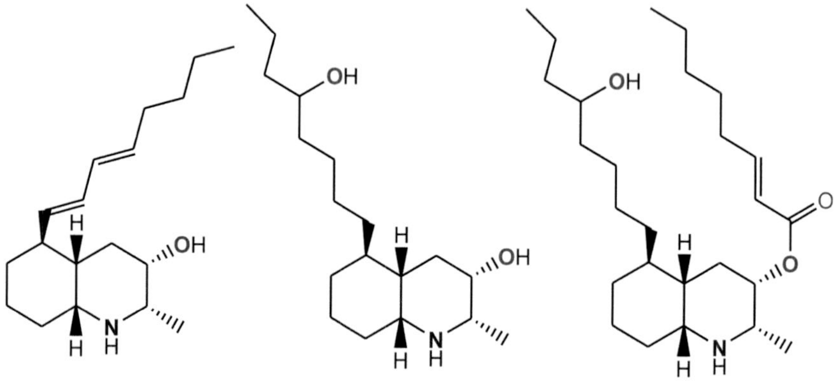

52
Lepadin -E

53
Lepadin-B

54
Lepadin-D

FIGURE 1.20 Chemical structures of Lepadins E, B, and D.

1.6.4.1 MoA OF ALKALOIDS

Though it is essential to understand the structural complexity and MoA to develop potent and/or structurally simplified analogs which can exhibit antimalarial activity, marine alkaloids are poorly studied compounds and not enough is known of them. Few reports have mentioned wherein alkaloids like Oroidin inhibited enoyl-ACP reductase, an enzyme involved in the parasite fatty acid biosynthesis thereby killing *P. falciparum* (Tasdemir et al., 2007). Similarly, salinosporamide A, a lactam alkaloid isolated from a marine bacterium of the new genus *Salinispora*, acts as a parasite proteasome inhibitor and possesses a good antimalarial activity (Feling et al., 2003).

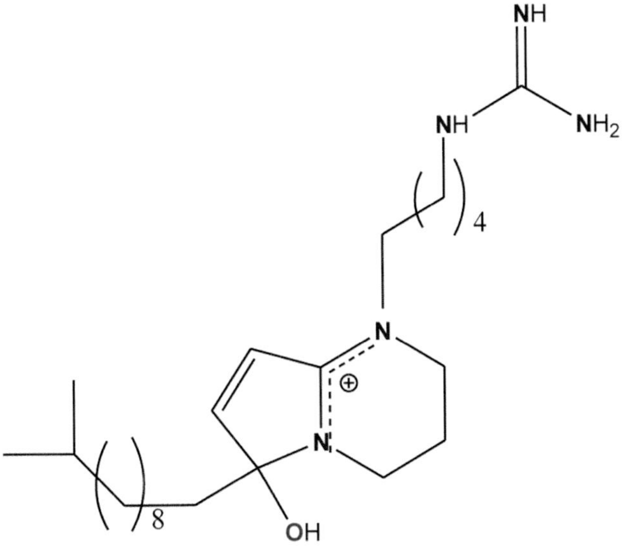

55

Phloeodictyn

FIGURE 1.21 Chemical structure of the active compound Phloeodictyn.

1.6.5 NONALKALOIDS

Antimalarials are not just limited to an isonitrile, endoperoxide, or alkaloids. They can be further categorized into quinones, phenols, and peptides (Fattorusso and Taglialatela-Scafati, 2009). Plants are also rich source of nonalkaloidal natural compounds which have antiplasmodial and antimalarial properties. They are classified into classes of terpenes, flavonoids, xanthones, chromones, anthraquinones, limonoids, miscellaneous, and related compounds.

1.6.5.1 NONALKALOIDS AS ANTIMALARIALS FROM PLANT PRODUCTS

1.6.5.1.1 Terpenes and Related Compounds

Sesquiterpene lactones were isolated from acetone extract of *Distephanus angulifolius* (Asteraceae) and proved to be the principle antiplasmodial element of the isolated compounds with IC50 values in the range 1.63.8 μM and 2.14.9 μM against chloroquine-sensitive D10 and chloroquine-resistant W2 *P. falciparum* strains, respectively (Fig. 1.22) (Pedersen et al., 2009).

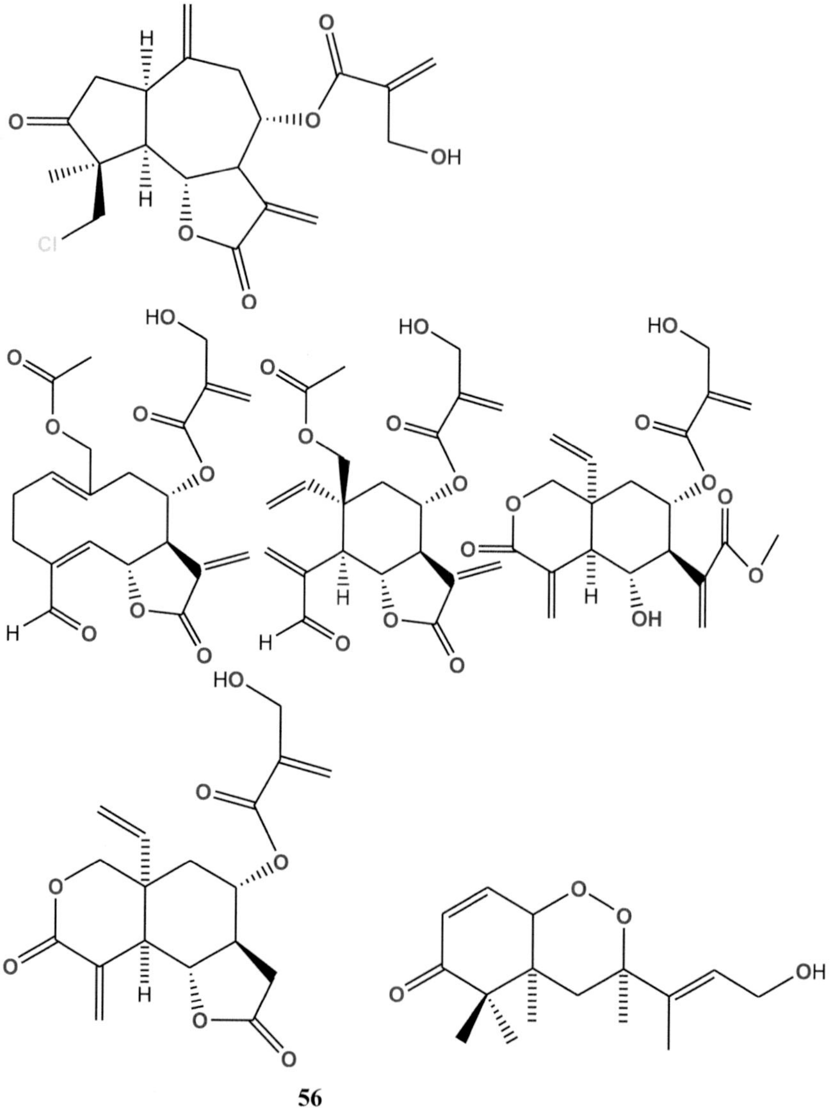

56

FIGURE 1.22 Structure of various terpenes and related compounds.

Bark of *Ekebergia capensis* produce 10 new triterpenoid compounds named ekeberins A and these compounds exhibit good antiplasmodial activity when screened *in vitro* against both chloroquine-sensitive (FCR-3) and -resistant (K-1) *P. falciparum* isolates. Compounds mainly (7-deacetoxy-7-oxogedunin) and (2-hydroxymethyl-2,3,22,23-tetrahydroxy-2,6,10,15,19,23-hexamethyl-6,10,14,18-tetracosatetraene) showed IC50 values of 6 and 7 M, respectively (Fig. 1.23) (Murata et al., 2008).

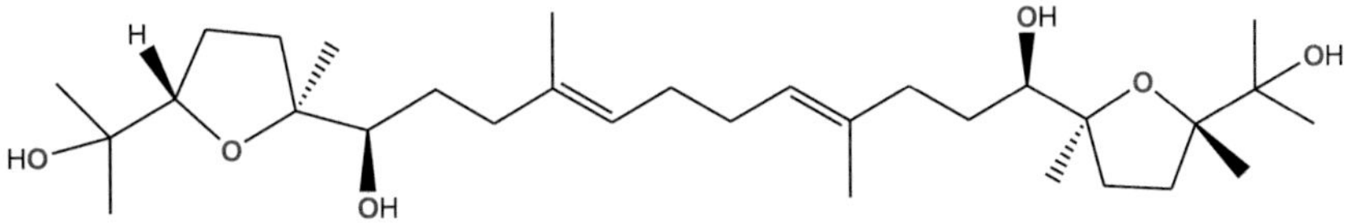

57

FIGURE 1.23 Structure of Ekeberins A (7-deacetoxy-7-oxogedunin) and (2-hydroxymethyl-2,3,22,23-tetrahydroxy-2,6,10,15,19,23-hexamethyl-6,10,14,18-tetracosatetraene.

Ethanol extract of the leaves of *Croton steenkampianus*, Gerstner (Euphorbiaceae) is the source of a new indanone derivative and two new diterpenoids (Fig. 1.24), and showed good results against chloroquine-susceptible strains (D10 and D6) and the chloroquine-resistant strains (Dd2 and W2) of *P. falciparum* at 9.1–15.8 M concentration (Fig. 1.24) (Adelekan et al., 2008). Similarly, two new cassane-type diterpenes (60, 61, 62) isolated from the seeds of *Bowdichia nitida* (Fabaceae) showed promising *in vitro* antiplasmodial activity against parasite *P. falciparum* 3D7 (IC50 = 1 M) and a good selectivity index with regard to the cytotoxicity on COLO201 cells (IC50 > 250 M) (Jonville et al ., 2008).

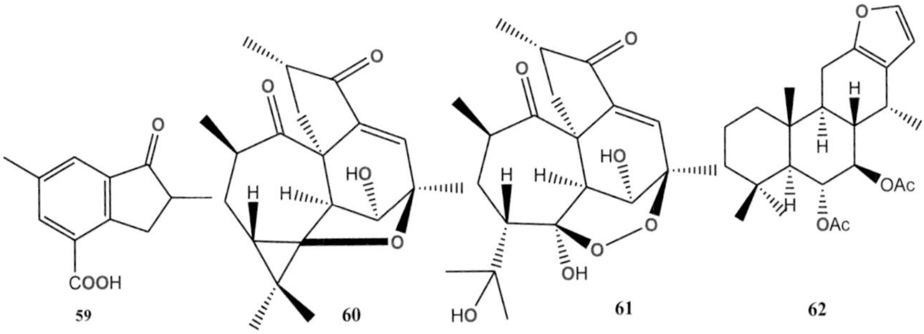

59 60 61 62

FIGURE 1.24 A new indanone derivative (**59**) and three new diterpenoids (**60, 61, 62**).

Isolation of three novel furanoterpenoids (Fig. 1.25) was done from the crude ethyl acetate extract of the medicinal plant, *Siphonochilus aethiopicus* (Zingiberaceae). Comparable in-vivo parasitemia reduction as well very good in-vitro activity was observed against the chloroquine-sensitive (D10) and chloroquine-resistant (K1) strains of *P. falciparum* with IC50 values of 2.9 and 1.4 g/mL, respectively (Lategan et al., 2009).

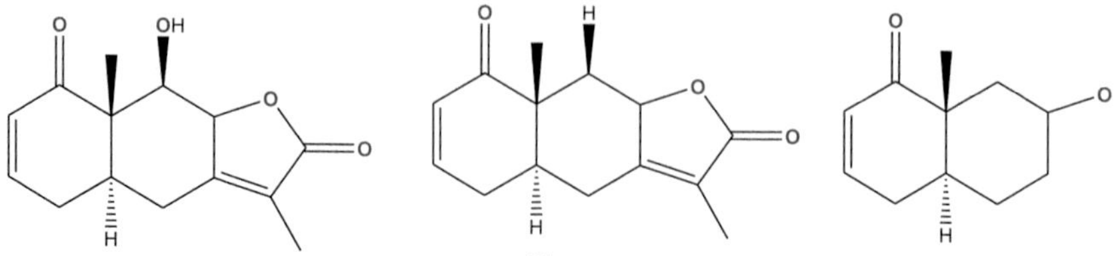

63 64 65

FIGURE 1.25 Structure of three isolated novel furanoterpenoids.

In-vitro assessment of pure natural monoterpenes geraniol, (−)-linalool, (−)-perillyl alcohol, (−)-isopulegol, (−)-limonene, and (±)-citronellol was done for their antiplasmodial activities against the chloroquine-resistant FcM29-Cameroon strain of *P. falciparum*. Only the alkyne (IC50 = 39 M) and cyclopentenone (IC50 = 1.8 M) derivatives showed a 13- and 290-fold increase in activity as compared to their parent molecules (Fig. 1.26). Based on these observations, new monoterpene derivatives should be harnessed like extracts from the stem bark of *Endodesmia calophylloides* (Guttiferae) showed potent antiplasmodial activities (Olagnier et al., 2007). Ethyl acetate extracts of novel friedelane triterpenoid named endodesmiadiol, friedelin, canophyllal, canophyllol, cerin, morelloflavone, volkensiflavone, 8-deoxygartanin, 3-acetoxyoleanolic acid, and 1,8-dihydroxy-3-isoprenyloxy-6-methylxanthone (Fig. 1.26), also exhibited antimalarial activity at IC50 values ranging from 7.2 to 23.6 μM (Ngouamegne et al., 2008).

Methanolic extract of *Carpesium rosulatum* (Asteraceae) yields Ineupatorolide A and have in-vitro antiplasmodial activity (IC50 = 8.2 g/mL) against the chloroquine-resistant D10 strain of *P. falciparum* (Fig. 1.27) (Moon, 2007). In addition, it has been noted that structural changes in the molecule significantly alter the anti-*Plasmodium* activity, thus betulinic acid and its derivative compounds might be considered as potential lead compounds for the development of new antimalarial drugs (de Sa et al., 2009).

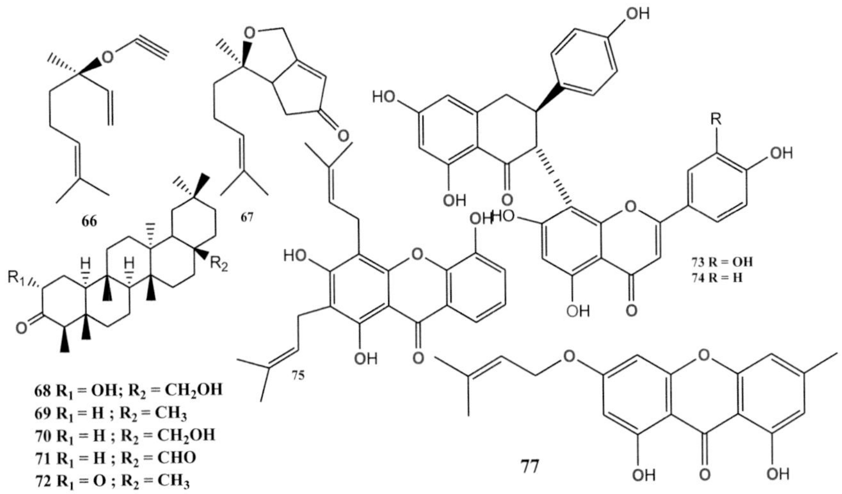

FIGURE 1.26 Structure of alkyne (IC50 = 39 μM) (**66**) and cyclopentenone (IC50 = 1.8 μM) (**67**) derivatives of monoterpenes. Friedelane triterpenoid, endodesmiadiol (**68**), friedelin (**69**), canophyllol (**70**), canophyllal (**71**), cerin (**72**), morelloflavone (**73**), volkensiflavone (**74**), 8-deoxygartanin (**75**), 3β-acetoxyoleanolic acid (**76**), and 1,8-dihydroxy-3-isoprenyloxy-6-methylxanthone (**77**).

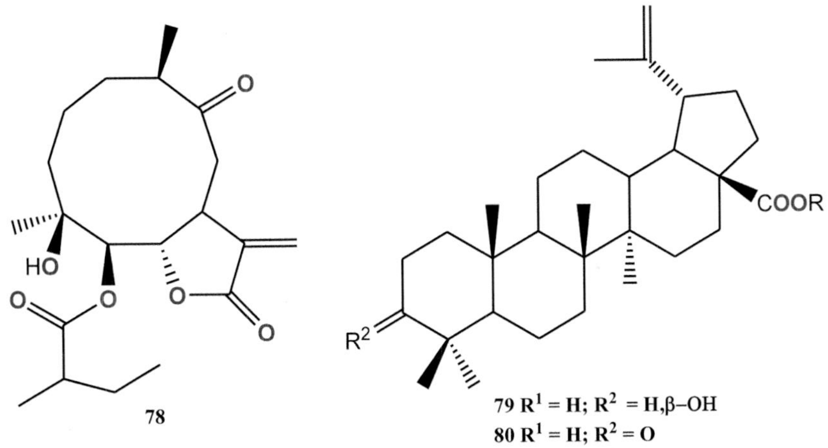

79 R^1 = H; R^2 = H,β–OH
80 R^1 = H; R^2 = O
81 R^1 = H; R^2 = H, β–OAc
82 R^1 = CH_3; R^2 = H, β–OH
83 R^1 = CH_3; R^2 = H, β–OAc

FIGURE 1.27 Structure of Ineupatorolide A (**78**) and betulinic acid and its derivative compounds (**79–83**).

Secondary metabolites from *Meliaceae* species and crude ethanol extract from the bark of *Chisocheton ceramicus* yielded Limonoids (Fig. 1.28) which showed a potent in-vitro antiplasmodial activity against *P. falciparum* 3D7 (IC50 = 0.56 μM). The presence of a tetrahydrofuran ring at C-4/C-6 and C-28 in limonoids is the reason for its higher antiplasmodial (Mohamad et al., 2009). Similarly, three triterpenes have been isolated from *Ogniauxia podolaena* (Cucurbitaceae), namely, cucurbitacin B, cucurbitacin D, and 20-epibryonolic acid (Fig. 1.28), which exhibited good antiplasmodial activity against the chloroquine-resistant FcM29 strain of *P. falciparum*, displaying IC50 values of 2.9, 7.7, and 4.3 μM, respectively.

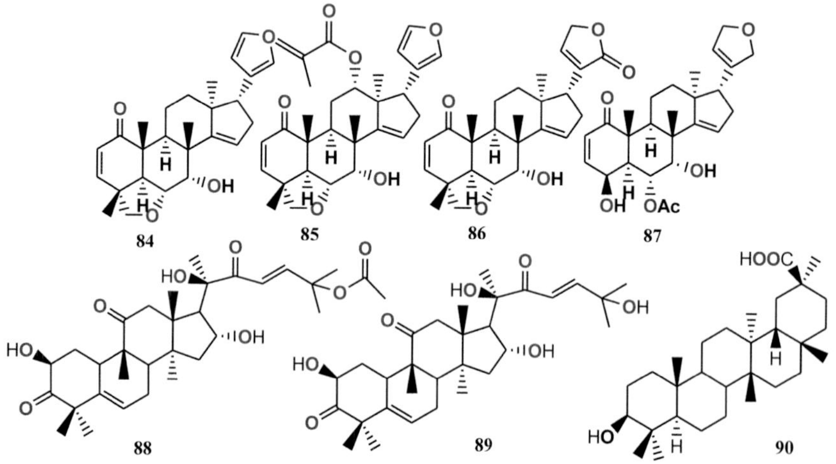

FIGURE 1.28 Structure of limonoids, ceramicines B–D (**84**–**86**), ceramicine A (**87**), cucurbitacin B (**88**), cucurbitacin D (**89**), and 20-epibryonolic acid (**90**).

1.6.5.1.2 Flavonoids, Quinones, and Related Compounds

In addition, flavonoids like obovatachalcone, obovatin, obovatin, guelin, a new -hydroxydihydrochalcone named (*S*)-elatadihydrochalcone (Fig. 1.29), derived from *Tephrosia elata*, was known to possess antiplasmodial activity as it was shown that -hydroxydihydrochalcones has an assigned -hydroxydihydrochalcone skeleton which is responsible for this activity (Muiva et al., 2009).

In Indonesia, a traditional medicine is used in the treatment of periodic fever and malaria. It was noted that a new bischromone named chrobisiamone A from the leaves of *Cassia siamea*, a *Fabaceae* species, bears good in-vitro antiplasmodial activity against parasite *P. falciparum* 3D7 (IC50 = 5.6 μM) (Fig. 1.30) (Oshimi et al., 2008).

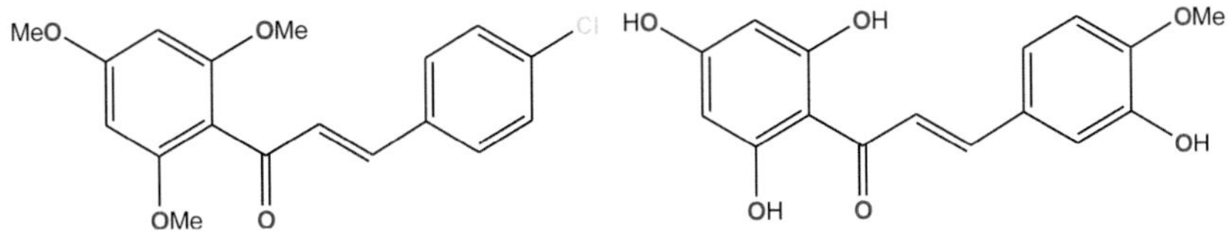

91 **92**

FIGURE 1.29 Structure of β-hydroxydihydrochalcone named (S)-elatadihydrochalcone.

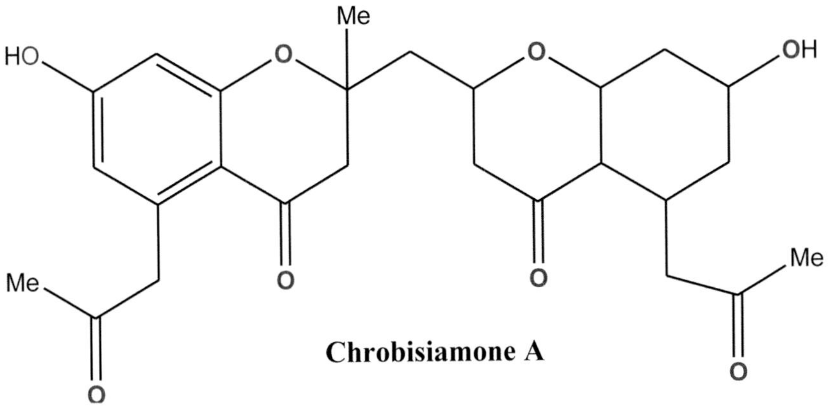

93

FIGURE 1.30 Structure of chrobisiamone A.

Appropriate insertion of prenyl groups into flavonoids often lead to more useful derivatives for construction of an antimalarial agent. Thus, flavonoids isolated from the stem bark of *Erythrina fusca* Lour., namely, lupinifolin, citflavanone, erythrisenegalone, lonchocarpol A, liquiritigenin, and 8-prenyldaidzein (Fig. 1.31) showed significant antiplasmodial activity (IC50 = 7.5 μg/mL) (Khaomek et al., 2008).

Xanthones were also shown to exhibit good antiplasmodial activity with IC50 values ranging from 4.4 to 8.0 μM against the *P. falciparum* chloroquine-resistant FcB1 strain. Four new xanthones named butyraxanthones A–D (Fig. 1.32; 100–103) and six known xanthones and a triterpenoid (lupeol) (Fig. 1.32; 104–109) are isolated from stem bark of *Pentadesma butyracea* Sabine (Clusiaceae) (Fig. 1.31) (Zelefack et al., 2009).

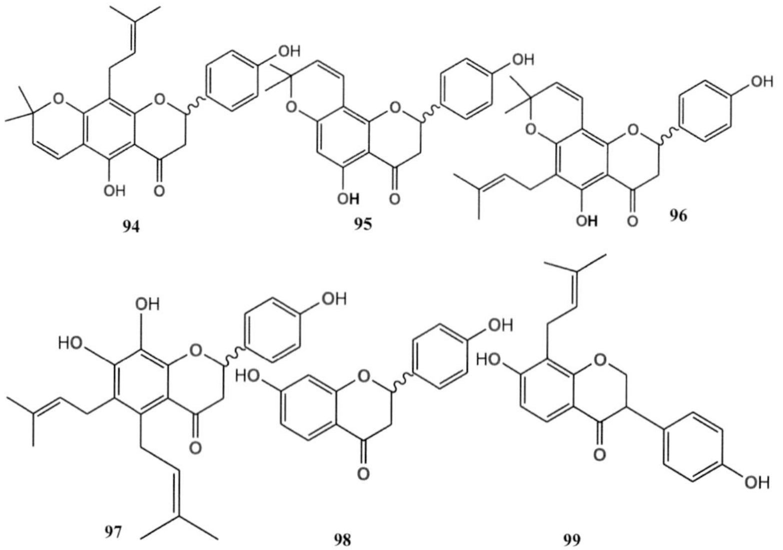

FIGURE 1.31 Structure of the flavonoids Lupinifolin (**94**), Citflavanone (**95**), Erythrisenegalone (**96**), Lonchocarpol A (**97**), Liquiritigenin (**98**), and 8-prenyldaidzein (**99**).

1.6.5.1.2.1 MoA of Flavonoids

Most of the flavonoids, like licochalcone A, preferentially inhibit the bc1 complex of *P. falciparum* via new permeation pathways (Ziegler, et al., 2004). Luteolin, the common dietary flavonoids, did not affect parasite susceptibility to the antimalarial drugs chloroquine or artemisinin with IC50 values of 11 ± 1 μM and 12 ± 1 μM for 3D7 and 7G8, respectively, but was found to prevent the progression of parasite growth beyond the young trophozoite stage. In addition, combining low concentrations of flavonoids showed an obvious additive antiplasmodial effect. Thus, flavonoid combinations as antiplasmodial agents need further investigation (Lehane and Saliba, 2008).

1.6.5.2 NONAKALOIDS AS ANTIMALARIALS FROM MARINE ORGANISMS

1.6.5.2.1 Quinones

Vanuatu marine sponge *Xestospongia* sp. produced Xestoquinone and Halenaquinone, which were tested for *P. falciparum* protein kinase inhibitory activity (Pfnek-1) and it showed antiplasmodial activity against an FCB1 *P. falciparum* strain but with a feeble selectivity index (SI 7) (Lebouvier et al., 2006; Laurent et al., 2006). Another Australian marine sponge *Dactylospongia elegans* also synthesized quinone, ilimaquinone (Fig. 1.33), which also showed antimalarial activity (Goclik et al., 2000).

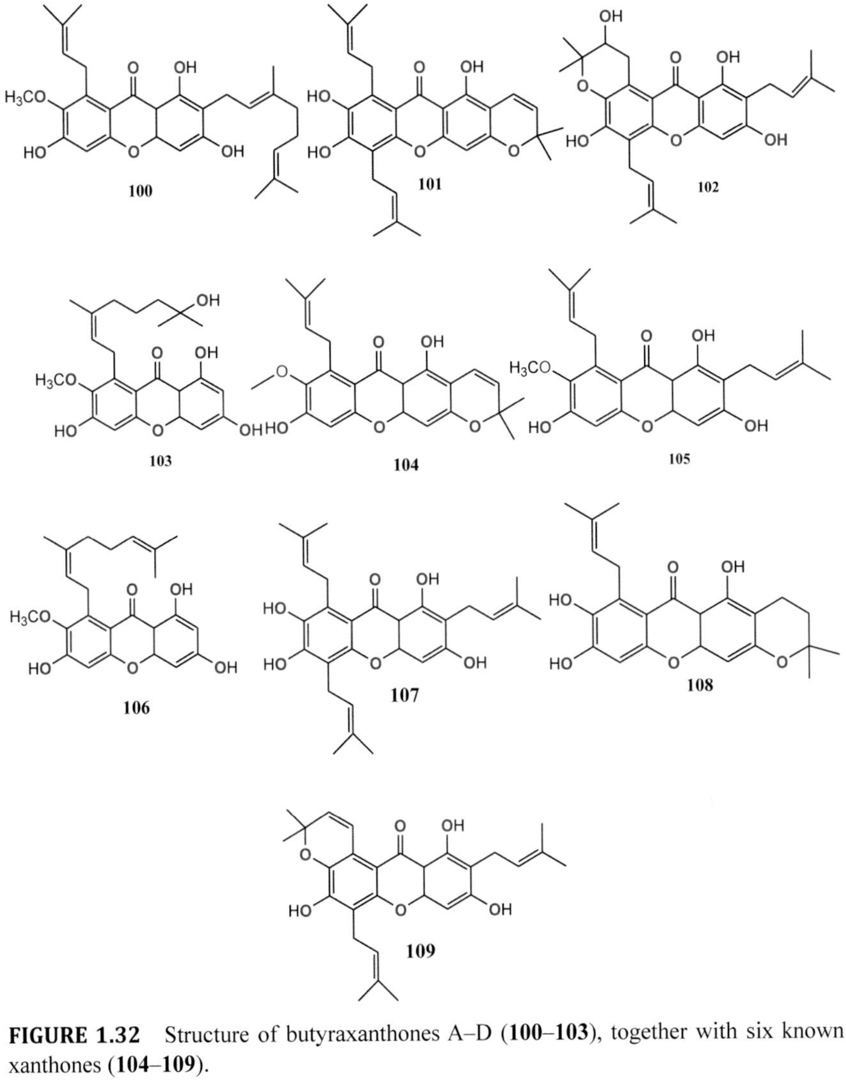

FIGURE 1.32 Structure of butyraxanthones A–D (**100–103**), together with six known xanthones (**104–109**).

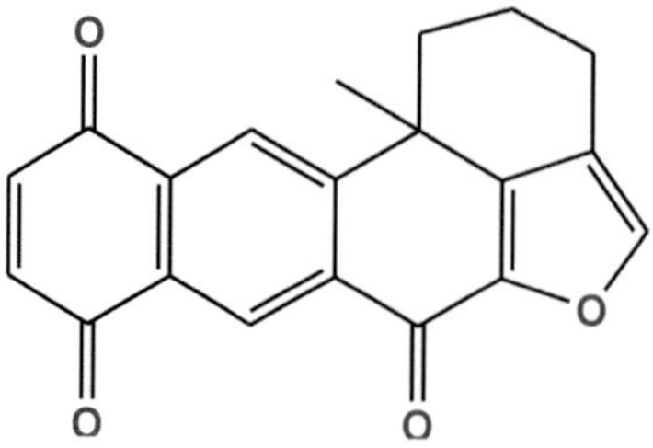

110
Xestoquinone

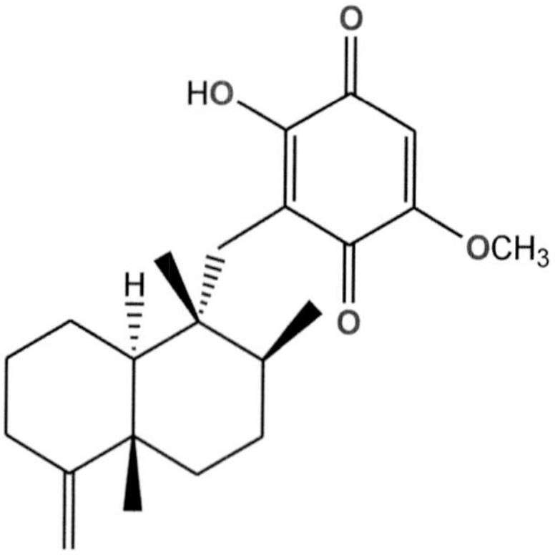

111
Ilimaquinone

FIGURE 1.33 Chemical structures of xestoquinone and ilimaquinone.

In parallel to xestoquinone, alisiaquinones (Fig. 1.34) showed not only structural similarity but also similar micromolar antiplasmodial activity against Pfnek-1. Nonetheless, alisiaquinone C which is formed by an addition of heterocycle and substitution of a taurine group increases the antiplasmodial activity to submicromolar levels (IC50 ~ 0.1 M) on *P. falciparum* with competitive selectivity index on the different plasmodial strains. This increase in activity is attributed to significant inhibition of the plasmodial enzyme farnesyl transferase and not to inhibition of Pfnek-1 a (Desoubzdanne et al., 2008).

1.6.5.2.2 *Phenols*

Representative of a distinctive family of sponge metabolites; 15-oxopuupehenol is a phenol-containing antimalarial marine metabolite isolated from sponges of the genus *Hyrtios*. It contains the quinol-quinone pair of avarol and avarone which aid in exhibiting in-vitro activity against *P. falciparum*. (*S*)-Curcuphenol is another sesquiterpene phenol isolated from different marine sponges belonging to genus *Didiscus coxeata*, which shows antimalarial activity (Fig. 1.35) (Nasu et al., 1995; El Sayed et al., 2001).

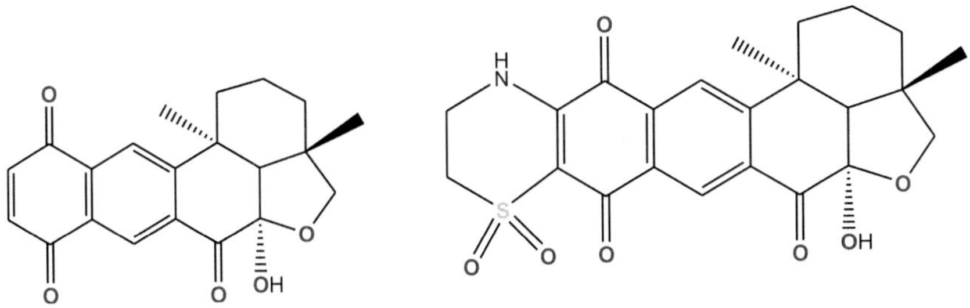

112
Alisiaquinones A

113
Alisiaquinones C

FIGURE 1.34 Chemical structure of alisiaquinones A and C.

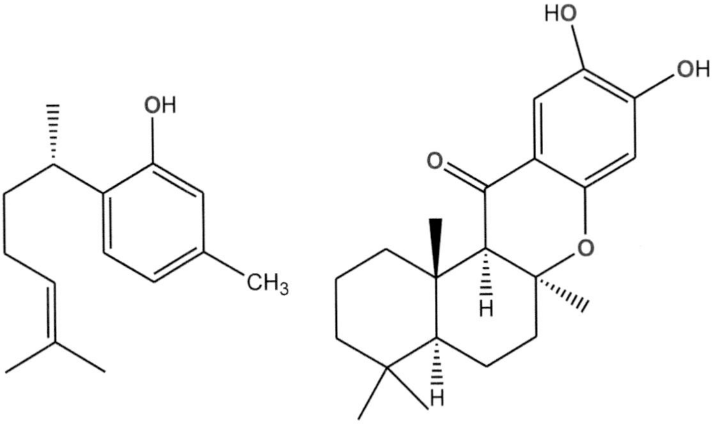

114

15-Oxopuupehenol

115

Cucurphenol

FIGURE 1.35 Chemical structures of 15-oxopuupehenol (**114**) and cucurphenol (**115**).

1.6.5.2.3 Peptides

Marine cyanobacterium *Oscillatoria* sp. and *Lyngbya majuscula* have proven to be a good source for cyclical (hexapeptides, venturamides) and linear alkynoiclipopeptide, respectively (Fig. 1.36). Both these compounds exhibited a moderate activity against *P. falciparum* (IC50 ~ 6–7 M), with a good selectivity. But a related nonaromatic dragonamide B, isolated from the same source, was completely devoid of activity. Another cyanobacterial peptide derivative, named gallinamide A, was isolated from *Schizothrix* species and contained unusual 4-(*S*)-amino-2-(*E*)-pentenoic acid subunit and a *N*,*N*-dimethyl isoleucine terminus. This compound also showed moderate in-vitro antimalarial activity (IC50 = 8.4 M) (Linington et al., 2008).

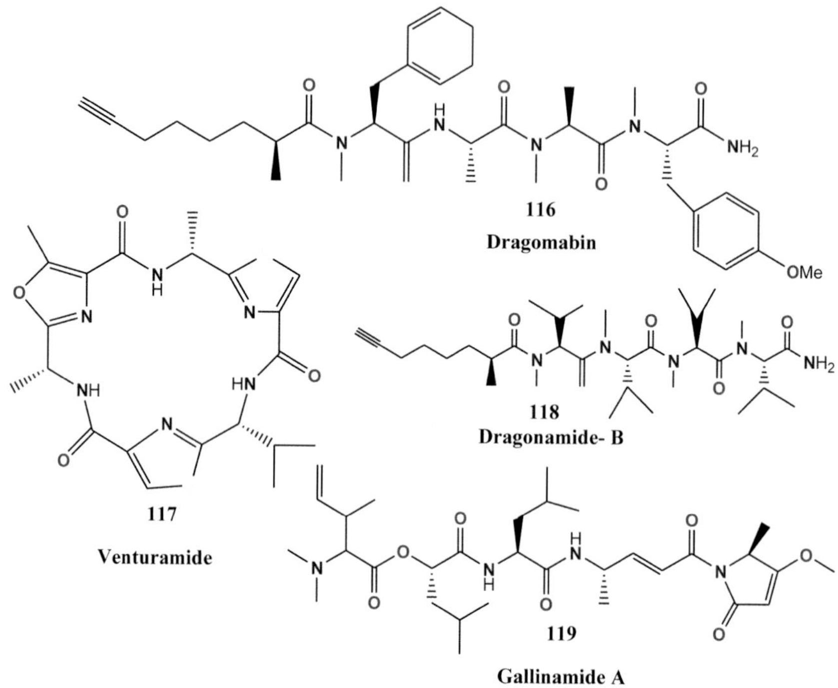

FIGURE 1.36 Chemical structures of dragomabin (**116**), venturamide A (**117**), dragonamide B (**118**), and gallinamide A (**119**).

1.6.5.2.4 Polyether

Polyether, a secondary metabolite, showed good selectivity toward in-vitro antimalarial activity (IC50 ~ 150 ng/mL). These compounds are lipophilic in nature and are believed to act as ionophores, and, thereby they are particularly able to interact with the protozoan-infected cell membrane (Fig. 1.37) (Otoguro et al., 2002, 2003; Na et al., 2008).

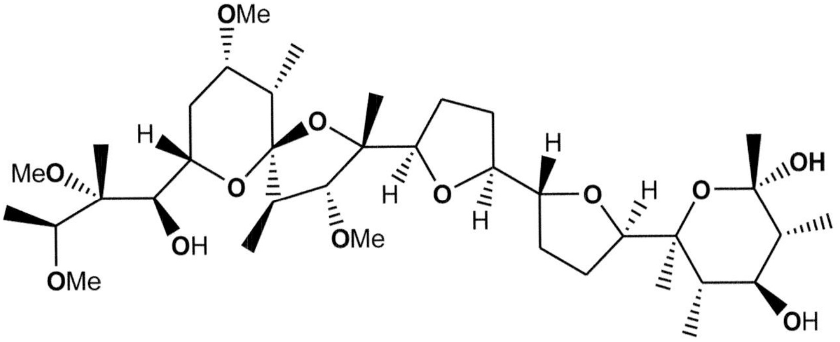

120

Polyether

FIGURE 1.37 Chemical structure of the polyether (**120**).

1.7 CONCLUSIONS

Lead compounds are derived from many sources and NPs have been the forefront in the treatment of VBDs. Though biologists, pharmacologists, and chemists all over world have unraveled quite a few lead compounds as potential drug targets for VBDs like malaria, leishmaniasis, yellow fever, and so on; this chapter is restricted to the structural and functional diversity of lead compounds as antimalarials. These lead compounds have greatly assisted to understand the MoA of various diseases particularly Malaria, thus helping in developing newer therapies via efficient and competent research. This has paved the way for evolving new and better drugs from natural bioresources using bioactive compounds as leads rather than the design of new compounds by de-novo synthesis. Although de novo synthesis may not be the favored route to new bioactive chemical entities, the chemical modification of functional groups on bioactive NP compounds offers huge capacity for drug development. This approach far exceeds the limitations of secondary metabolism and effectively complements the biosynthetic machinery with the entire scope of synthetic chemistry. This chapter aimed to demonstrate both plant- and marine-derived compounds that usually serve as leads for the development of new classes of antimalarial drugs. It is also stated that the drug potential varies in the toxicity, availability, and the chemical structure of the compounds.

But it raises a serious environmental concern. Overexploitation of marine biodiversity should be avoided and alternate mechanisms like bacterial fermentation combined with genetic engineering should be employed to increase the chances of their full pharmacological evaluation in possible therapies.

KEYWORDS

- **lead compounds**
- **antimalarials**
- **isonitrile derivatives**
- **alkaloids**
- **endoperoxide derivatives**
- **quinones**
- **phenols**
- **polyethers**
- **peptides**

REFERENCES

Adelekan, A. M.; Prozesky, E. A.; Hussein, A. A.; Urena, L. D.; van Rooyen, P. H.; Liles, D. C.; Meyer, J. J.; Rodriguez, B. Bioactive Diterpenes and Other Constituents of *Croton steenkampianus*. *J. Nat. Prod.* **2008,** *71*, 1919–1922.

Adeline, M. T.; Debitus, C. Phloeodictines A and B: New Antibacterial and Cytotoxic Bicyclic Amidinium Salts from the New Caledonian Sponge, *Phloeodictyon* sp. *J. Org. Chem.* **1992,** *57*, 3832–3835.

Agarwal, D.; Gupta, R. D.; Awasthi, S. K. Are Antimalarial Hybrid Molecules a Close Reality or a Distant Dream? *Antimicrob. Agent Chemother.* **2017,** *61* (5), e00249–e00217.

Amer, A.; Mehlhorn, H. Larvicidal Effects of Various Essential Oils against *Aedes*, *Anopheles*, and *Culex* larvae (Diptera, Culicidae). *Parasitol. Res.* **2006,** *99* (4), 466–472.

Ang, K. K. H.; Holmes, M. J.; Higa, T.; Hamann, M. T.; Kara, U. A. K. In Vivo Antimalarial Activity of the Beta-carboline Alkaloid Manzamine A. *Antimicrob. Agent Chemother.* **2000,** *44*, 1645–1649.

Angerhofer, C. K.; Pezzuto, J. M.; Koenig, G. M.; Wright, A. D.; Sticher, O. Antimalarial Activity of Sesquiterpenes from the Marine Sponge *Acanthella klethra*. *J. Nat. Prod.* **1992,** *55*, 1787–1789.

Angerhofer, C. K.; Guinaudeau, H.; Wongpanich, V.; Pezzuto, J. M.; Cordell, G. A. Antiplasmodial and Cytotoxic Activity of Natural Bisbenzylisoquinoline Alkaloids. *J Nat. Prod.* **1999,** *62* (1), 59–66.

Antonova-Koch, Y.; Meister, S.; Abraham, M.; Luth, M. R.; Ottilie, S.; Lukens, A. K.; Sakata-Kato, T.; Vanaerschot, M.; Owen, E.; Jado, J. C.; Maher, S. Open-Source Discovery of Chemical Leads for Next-Generation Chemoprotective Antimalarials. *Science* **2018,** *362* (6419), eaat9446.

Ashley, E, A.; Phyo, A. P. Drugs in Development for Malaria. *Drugs* **2018,** *78* (9), 861–879.

Avery, M. A.; Chong, W. K. M.; Jennings-White, C. Stereoselective Total Synthesis of (+)-Artemisinin, the Antimalarial Constituent of *Artemisia annua* L. *J. Am. Chem. Soc.* **1992,** *114*, 974–979.

Basco, L. K.; Ramiliarisoa, O.; Le Bras, J. *In Vitro* Activity of Pyrimethamine, Cycloguanil, and Other Antimalarial Drugs against African Isolates and Clones of *Plasmodium falciparum*. *Am. J. Trop. Med. Hyg.* **1994,** *50* (2), 193–199.

Bazala, V. *The Historical Development of Medicine in the Croatian Lands*; Croatian Publishing Bibliographic Institute: Zagreb, 1943; pp 9–20.

Bojadzievski, P. *The Health Services in Bitola through the Centuries*; Society of Science and Art: Bitola, 1992; pp 15–27.

Bringmann, G.; Feineis, D. Novel Antiparasitic Biaryl Alkaloids from West African Dioncophyllaceae Plants. *Actual. Chim. Therap.* **2000,** *26*, 151–172.

Bringmann, G.; Rummey, C. 3D QSAR Investigations on Antimalarial Naphthylisoquinoline Alkaloids by Comparative Molecular Similarity Indices Analysis (CoMSIA), Based on Different Alignment Approaches. *J. Chem. Inform. Comp. Sci.* **2003,** *43* (1), 304–316.

Bringmann, G.; Wohlfarth, M.; Rischer, H.; Heubes, M.; Saeb, W.; Diem, S.; Herderich, M.; Schlauer, J. A Photometric Screening Method for Dimeric Naphthylisoquinoline Alkaloids and Complete On-line Structural Elucidation of a Dimer in Crude Plant Extracts, by the LC–MS/LC–NMR/LC–CD Triad. *Anal. Chem.* **2001,** *73* (11), 2571–2577.

Bueno-Marí, R. Looking for New Strategies to Fight against Mosquito-Borne Diseases: Toward the Development of Natural Extracts for Mosquito Control and Reduction of Mosquito Vector Competence. *Front. Physiol.* **2013,** *13*, 4–20.

Butler, M. S. The Role of Natural Product Chemistry in Drug Discovery. *J. Nat. Prod.* **2004,** *67* (12), 2141–2153.

Butler, M. S. Natural Products to Drugs: Natural Product-Derived Compounds in Clinical Trials. *Nat. Prod. Rep.* **2008,** *25* (3), 475–516.

Carraz, M.; Jossang, A.; Franetich, J. F.; Siau, A.; Ciceron, L.; Hannoun, L.; Sauerwein, R.; Frappier, F.; Rasoanaivo, P.; Snounou, G.; Mazier, D. A Plant-Derived Morphinan as a Novel Lead Compound Active against Malaria Liver Stages. *PLoS Med.* **2006,** *3* (12), e513.

Cimanga, K.; De Bruyne, T.; Pieters, L.; Vlietinck, A. J.; Turger, C. A. *In Vitro* and *In Vivo* Antiplasmodial Activity of Cryptolepine and Related Alkaloids from *Cryptolepis sanguinolenta*. *J. Nat. Prod.* **1997,** *60* (7), 688–691.

Cragg, G. M.; Kingston, D. G.; Newman, D. J. *Anticancer Agents from Natural Products*; CRC Press: Boca Raton, FL, 2011; p 10.

Cragg, G. M.; Katz, F.; Newman, D. J.; Rosenthal, J. The Impact of the United Nations Convention on Biological Diversity on Natural Products Research. *Nat. Prod. Rep.* **2012,** *29* (12), 1407–1423.

D'Ambrosio, M.; Guerriero, A.; Deharo, E.; Debitus, C.; Munoz, V.; Pietra, F. New Types of Potentially Antimalarial Agents: Epidioxy-Substituted Norditerpene and Norsesterterpenes from the Marine Sponge *Diacarnus levii*. *Helv. Chim. Acta* **1998,** *81*, 1285–1292.

David, A. J.; David, T. Y. Strategies to Enhance Medical Counter Measures after the Use of Chemical Warfare Agents on Civilians. *Handbook of Toxicology of Chemical Warfare Agents*; Academic Press: Cambridge, MA, 2009.

de Sa, M. S.; Costa, J. F.; Krettli, A. U.; Zalis, M. G.; Maia, G. L.; Sette, I. M.; Camara, C. D.; Filho, J. M.; Giulietti-Harley, A. M.; Ribeiro Dos Santos, R.; Soares, M. B. Antimalarial Activity of Betulinic Acid and Derivatives *In Vitro* against *Plasmodium falciparum* and *In Vivo* in *P. berghei* Infected Mice. *Parasitol. Res.* **2009,** *105*, 275–279.

Delves, M.; Plouffe, D.; Scheurer, C.; Meister, S.; Wittlin, S.; Winzeler, E. A.; Sinden, R. E.; Leroy, D. The Activities of Current Antimalarial Drugs on the Life Cycle Stages of *Plasmodium*: A Comparative Study with Human and Rodent Parasites. *PLoS Med.* **2012,** *9* (2), e1001169.

Dervendzi, V. *Contemporary Treatment with Medicinal Plants*; Tabernakul: Skopje, 1992; pp 5–43.

Desoubzdanne, D.; Marcourt, L.; Raux, R.; Chevalley, S.; Dorin, D.; Doerig, C.; Valentin, A.; Ausseil, F.; Debitus, C. Alisiaquinones and Alisiaquinol, Dual Inhibitors of *Plasmodium falciparum* Enzyme Targets from a New Caledonian Deep Water Sponge. *J. Nat. Prod.* **2008,** *71* (7), 1189–1192.

Dev, S. Ancient Modern Concordance in Ayurvedic Plants: Some Examples. *Environ. Health Perspect.* **1999,** *107* (10), 783.

Di Blasio, B.; Fattorusso, E.; Magno, S.; Mayol, L.; Pedone, C.; Santacroce, C.; Sica, D. Axisonitrile-3, Axisothiocyanate-3 and Axamide-3: Sesquiterpenes with a Novel Spiro[4,5] decane Skeleton from the Sponge *Axinella cannabina*. *Tetrahedron* **1976,** *32*, 473–478.

Dimitrova, Z. *The History of Pharmacy*; St. Clement of Ohrid: Sofija, 1999; pp 13–26.

Eckstein-Ludwig, U.; Webb, R. J.; Van Goethem, I. D. A.; East, J. M.; Lee, A. G.; Kimura, M.; O'Neill, P. M.; Bray, P. G.; Ward, S. A.; Krishna, S. Artemisinins Target the SERCA of *Plasmodium falciparum*. *Nature* **2003,** *424*, 957–961.

El Sayed, K. A.; Kelly, M.; Kara, U. A. K.; Ang, K. K. H.; Katsuyama, I.; Dunbar, D. C.; Khan, A. A.; Hamann, M. T. New Manzamine Alkaloids with Potent Activity against Infectious Diseases. *J. Am. Chem. Soc.* **2001,** *123*, 1804–1808.

Elujoba, A. A. Traditional Medicinal Plants and Malaria. *Afr. J. Trad., Compl. Alternat. Med.* **2005,** *2* (2), 206–207.

Farnsworth, N. R.; Akerele, O.; Bingel, A. S.; Soejarto, D. D.; Guo, Z. Medicinal Plants in Therapy. *Bull. World Health Organ.* **1985,** *63*, 965–981.

Fattorusso, E.; Taglialatela-Scafati, O. Marine Antimalarials. *Mar. Drugs* **2009,** *7* (2), 130–152.

Fattorusso, C.; Campiani, G.; Catalanotti B.; Persico, M.; Basilico, N.; Parapini, S.; Taramelli, D.; Campagnuolo, C.; Fattorusso, E.; Romano, A.; Taglialatela-Scafati, O. Endoperoxide Derivatives from Marine Organisms: 1,2-Dioxanes of the Plakortin Family as Novel Antimalarial Agents. *J. Med. Chem.* **2006,** *49* (24), 7088–7094.

Feling, R. H.; Buchanan, G. O.; Mincer, T. J.; Kauffman, C. A.; Jensen, P. R.; Fenical, W. Salinosporamide A: A Highly Cytotoxic Proteasome Inhibitor from a Novel Microbial Source, a Marine Bacterium of the New Genus *Salinispora*. *Angew. Chem., Int. Ed.* **2003,** *42*, 355–357.

Fotie, J. Key Natural Products in Malaria Chemotherapy: From Quinine to Artemisinin and Beyond. *Bioactive Natural Products: Opportunities and Challenges in Medicinal Chemistry*; World Scientific: Singapore, 2012; pp 223–271.

Frederich, M.; Tits, M.; Angenot, L. Potential Antimalarial Activity of Indole Alkaloids. *Trans. R. Soc. Trop. Med. Hyg.* **2008,** *102* (1), 11–19.

Goclik, E.; König, G. M.; Wright, A. D.; Kaminsky, R. Pelorol from the Tropical Marine Sponge *Dactylospongia elegans*. *J. Nat. Prod.* **2000,** *63* (8), 1150–1152.

Gorunovic, M.; Lukic, P. Pharmacognosy. *Beograd: Gorunovic M* **2001,** 1–5.

Guantai, E.; Chibale, K. How Can Natural Products Serve as a Viable Source of Lead Compounds for the Development of New/Novel Antimalarials? *Mal. J.* **2011,** *10* (1), S2.

Hemingway, J.; Hawkes, N. J.; McCarroll, L.; Ranson, H. The Molecular Basis of Insecticide Resistance in Mosquitoes. *Ins. Biochem. Mol. Biol.* **2004,** *34* (7), 653–665.

Hien, T. T.; White, N. J. Qinghaosu. *Lancet* **1993,** *341* (8845), 603–608.

Huxtable, R. J.; Schwarz, S. K. The Isolation of Morphine—First Principles in Science and Ethics. *Mol. Interv.* **2001,** *1* (4), 189.

Jacquemond-Collet, I.; Benoit-Vical, F.; Valentin, M.; Stanislas, A.; Mallié, E.; Fourasté, M. Antiplasmodial and Cytotoxic Activity of Galipinine and Other Tetrahydroquinolines from *Galipea officinalis*. *Planta Med.* **2002,** *68* (01), 68–69.

Jonckers, T. H.; Van Miert, S.; Cimanga, K.; Bailly, C.; Colson, P.; De Pauw-Gillet, M. C.; van den Heuvel, H.; Claeys, M.; Lemière, F.; Esmans, E. L.; Rozenski, J. Synthesis, Cytotoxicity, and Antiplasmodial and Antitrypanosomal Activity of New Neocryptolepine Derivatives. *J. Med. Chem.* **2002,** *45* (16), 3497–3508.

Jones, K. E.; Patel, N. G.; Levy, M. A.; Storeygard, A.; Balk, D.; Gittleman, J. L.; Daszak, P. Global Trends in Emerging Infectious Diseases. *Nature* **2008,** *451* (7181), 990.

Jonville, M. C.; Capel, M.; Frederich, M.; Angenot, L.; Dive, G.; Faure, R.; Azas, N.; Ollivier, E. Fagraldehyde, A Secoiridoid Isolated from *Fagraea fragrans*. *J. Nat. Prod.* **2008,** *71*, 2038–2040.

Kamchonwongpaisan, S.; Meshnick, S. R. The Mode of Action of the Antimalarial Artemisinin and Its Derivatives. *Gen. Pharm. J.* **1996,** *27* (4), 587–592.

Kapoor, L. D. *Handbook of Ayurvedic Medicinal Plants: Herbal Reference Library*; Routledge: Abingdon, United Kingdom, 2017.

Kelly, K. *The History of Medicine: Early Civilizations: Prehistoric Times to 500 CE*; Facts on File: New York, NY, 2009.

Khaomek, P.; Ichino, C.; Ishiyama, A.; Sekiguchi, H.; Namatame, M.; Ruangrungsi, N.; Saifah, E.; Kiyohara, H.; Otoguro, K.; Omura, S.; Yamada, H. *In Vitro* Antimalarial Activity of Prenylated Flavonoids from *Erythrina fusca*. *J. Nat. Med.* **2008,** *62*, 217–220.

Koehn, F. E.; Carter, G. T. The Evolving Role of Natural Products in Drug Discovery. *Nat. Rev. Drug Disc.* **2005,** *4* (3), 206.

Kovacevic, N. *Fundamentals of Pharmacognosy*; Personal Edition: Beograd, 2000; pp 170–171.

Lategan, C. A.; Campbell, W. E.; Seaman, T.; Smith, P. J. The Bioactivity of Novel Furanoterpenoids Isolated from *Siphonochilus aethiopicus*. *J. Ethnopharmacol.* **2009,** *121*, 92–97.

Laurent, D.; Jullian, V.; Parenty, A.; Knibiehler, M.; Dorin, D.; Schmitt, S.; Lozach, O.; Lebouvier, N.; Frostin, M.; Alby, F.; Maurel, S. Antimalarial Potential of Xestoquinone, a Protein Kinase Inhibitor Isolated from a Vanuatu Marine Sponge *Xestospongia* sp. *Bioorg. Med. Chem.* **2006,** *14* (13), 4477–4482.

Lebouvier, N.; Frostin, M.; Alby, F.; Maurel, S.; Doerig, C.; Meijer, L.; Sauvain, M. Antimalarial Potential of Xestoquinone, a Protein Kinase Inhibitor Isolated from a Vanuatu Marine Sponge *Xestospongia* sp. *Bioorg. Med. Chem.* **2006,** *14*, 4477–4482.

Lee, A. H.; Symington, L. S.; Fidock, D. A. DNA Repair Mechanisms and Their Biological Roles in the Malaria Parasite *Plasmodium falciparum*. *Microbiol. Mol. Biol. Rev.* **2014,** *78* (3), 469–486.

Lee, K. H. Current Developments in the Discovery and Design of New Drug Candidates from Plant Natural Product Leads. *J. Nat. Prod.* **2004,** *67* (2), 273–283.

Lee, K. H. Novel Antitumor Agents from Higher Plants. *Med. Res. Rev.* **1999,** *9* (6), 569–596.

Lehane, A. M.; Saliba, K. J. Common Dietary Flavonoids Inhibit the Growth of the Intraerythrocytic Malaria Parasite. *BMC Res. Notes* **2008,** *1*, 26.

Li, G.; Hsung, R. P.; Slafer, B. W.; Sagamanova, I. K. Total Synthesis of (+)-Lepadin F. *Org. Lett.* **2008,** *10* (21), 4991–4994.

Lin, Z.; Hoult, J. R.; Bennett, D. C.; Raman, A. Stimulation of Mouse Melanocyte Proliferation by Piper Nigrum Fruit Extract and Its Main Alkaloid, Piperine. *Planta Med.* **1999,** *65* (7), 600–603.

Linington, R. G.; Clark, B. R.; Trimble, E. E.; Almanza, A.; Ureña, L. D.; Kyle, D. E.; Gerwick, W. H. Antimalarial Peptides from Marine Cyanobacteria: Isolation and Structural Elucidation of Gallinamide A. *J. Nat. Prod.* **2008,** *72* (1), 14–17.

Mancini, I.; Guella, G.; Sauvain, M.; Debitus, C.; Duigou, A.; Ausseil, F.; Menou, J.; Pietra, F. New 1,2,3,4-Tetrahydropyrrolo[1,2-*a*]Pyrimidinium Alkaloids (Phloeodictynes) from the New Caledonian Shallow-Water Haplosclerid Sponge *Oceanapia fistulosa*: Structural Elucidation from Mainly LC–Tandem-MS-Soft-Ionization Techniques and Discovery of Antiplasmodial Activity. *Org. Biomol. Chem.* **2004,** *2*, 783–787.

Mbengue, A.; Bhattacharjee, S.; Pandharkar, T.; Liu, H.; Estiu, G.; Stahelin, R. V.; Rizk, S. S.; Njimoh, D. L.; Ryan, Y.; Chotivanich, K.; Nguon, C. A Molecular Mechanism of Artemisinin Resistance in *Plasmodium falciparum* Malaria. *Nature* **2015,** *520* (7549), 683.

Mohamad, K.; Hirasawa, Y.; Litaudon, M.; Awang, K.; Hadi, A. H.; Takeya, K.; Ekasari, W.; Widyawaruyanti, A.; Zaini, N. C.; Morita, H. Ceramicines B–D, New Antiplasmodial Limonoids from *Chisocheton ceramicus*. *Bioorg. Med. Chem.* **2009,** *17*, 727–730.

Mok, S.; Ashley, E. A.; Ferreira, P. E.; Zhu, L.; Lin, Z.; Yeo, T.; Chotivanich, K.; Imwong, M.; Pukrittayakamee, S.; Dhorda, M.; Nguon, C. Population Transcriptomics of Human Malaria Parasites Reveals the Mechanism of Artemisinin Resistance. *Science* **2015,** *347* (6220), 431–435.

Moon, H. I. Antiplasmodial Activity of Ineupatorolides A from *Carpesium rosulatum*. *Parasitol. Res.* **2007,** *100*, 1147–1149.

Muiva, L. M.; Yenesew, A.; Derese, S.; Heydenreich, M.; Peter, M. G.; Akala, H. M.; Eyase, F.; Waters, N. C.; Mutai, C.; Keriko, J. M.; Walsh, D. Antiplasmodial β-Hydroxydihydrochalcone from Seedpods of *Tephrosia elata*. *Phytochem. Lett.* **2009,** *2* (3), 99–102. doi:10.1016/j.phytol. 2009.01.002.

Mukherjee, A. K.; Basu, S.; Sarkar, N.; Ghosh, A. C. Advances in Cancer Therapy with Plant Based Natural Products. *Curr. Med. Chem.* **2001,** *8* (12), 1467–1486.

Murata, T.; Miyase, T.; Muregi, F. W.; Naoshima-Ishibashi, Y.; Umehara, K.; Warashina, T.; Kanou, S.; Mkoji, G. M.; Terada, M.; Ishih, A. Antiplasmodial Triterpenoids from *Ekebergia capensis*. *J. Nat. Prod.* **2008,** *71*, 167–174.

Muriithi, M. W.; Abraham, W. R.; Addae-Kyereme, J.; Scowen, I.; Croft, S. L.; Gitu, P. M.; Kendrick, H.; Njagi, E. N.; Wright C. W. Isolation and *In Vitro* Antiplasmodial Activities of Alkaloids from *Teclea trichocarpa*: In Vivo Antimalarial Activity and X-ray Crystal Structure of Normelicopicine. *J. Nat. Prod.* **2002,** *65* (7), 956–959.

Na, M.; Meujo, D. A.; Kevin, D.; Hamann, M. T.; Anderson, M.; Hill, R. T. A New Antimalarial Polyether from a Marine *Streptomyces* sp. H668. *Tetrahedron Lett.* **2008,** *49* (44), 6282–6285.

Nasu, S. S.; Yeung, B. K.; Hamann, M. T.; Scheuer, P. J.; Kelly-Borges, M.; Goins, K. Puupehenone-Related Metabolites from Two Hawaiian Sponges, *Hyrtios* spp. *J. Org. Chem.* **1995,** *60* (22), 7290–7292.

Nevin, R. L.; Croft, A. M. Psychiatric Effects of Malaria and Anti-Malarial Drugs: Historical and Modern Perspectives. *Mal. J.* **2016,** *15* (1), 332.

Newman, D. J.; Cragg, G. M. Natural Products as Sources of New Drugs over the Last 25 years. *J. Nat. Prod.* **2007,** *70* (3), 461–477.

Newman, D. J.; Cragg, G. M. Natural Products as Sources of New Drugs from 1981 to 2014. *J. Nat. Prod.* **2016,** *79* (3), 629–661.

Newman, D. J.; Gordon, M. C. Natural Products as Sources of New Drugs over the 30 Years from 1981 to 2010. *J. Nat. Prod.* **2012,** *75* (3), 311–335.

Ngouamegne, E. T.; Fongang, R. S.; Ngouela, S.; Boyom, F. F.; Rohmer, M.; Tsamo, E.; Gut, J.; Rosenthal, P. J. Endodesmiadiol, a Friedelane Triterpenoid, and Other Antiplasmodial Compounds from *Endodesmia calophylloides*. *Chem. Pharm. Bull. (Tokyo)* **2008,** *56*, 374–377.

Nikolovski, B. *Essays on the History of Health Culture in Macedonia*; Macedonian Pharmaceutical Association: Skopje, 1995; pp 17–27.

Ojo, O. O.; Oluyege, J. O.; Famurewa, O. Antiviral Properties of Two Nigerian Plants. *Afr. J. Plant Sci.* **2009,** *3* (7), 157–159.

Olagnier, D.; Costes, P.; Berry, A.; Linas, M. D.; Urrutigoity, M.; Dechy-Cabaret, O.; Benoit-Vical, F. Modifications of the Chemical Structure of Terpenes in Antiplasmodial and Antifungal Drug Research. *Bioorg. Med. Chem. Lett.* **2007,** *17*, 6075–6078.

Oshimi, S.; Tomizawa, Y.; Hirasawa, Y.; Honda, T.; Ekasari, W.; Widyawaruyanti, A.; Rudyanto, M.; Indrayanto, G.; Zaini, N. C.; Morita, H. Chrobisiamone A, a New Bischromone from *Cassia siamea* and a Biomimetic Transformation of 5-Acetonyl-7-hydroxy-2-methylchromone into Cassiarin A. *Bioorg. Med. Chem. Lett.* **2008,** *18*, 3761–3763.

Otoguro, K.; Ishiyama, A.; Ui, H.; Kobayashi, M.; Manabe, C.; Yan, G.; Takahashi, Y.; Tanaka, H.; Yamada, H.; Omura, S. In Vitro and In Vivo Antimalarial Activities of the Monoglycoside Polyether Antibiotic, K-41 against Drug Resistant Strains of Plasmodia. *J. Antibiotics* **2002,** *55* (9), 832–834.

Otoguro, K.; Ui H.; Ishiyama, A.; Kobayashi, M.; Togashi, H.; Takahashi, Y.; Masuma, R.; Tanaka, H.; Tomoda, H.; Yamada, H.; Omura, S. *In Vitro* and *In Vivo* Antimalarial Activities of a Non-glycosidic 18-Membered Macrolide Antibiotic, Borrelidin, against Drug-Resistant Strains of Plasmodia. *J. Antibiotics* **2003,** *56* (8), 727–729.

Paterson, I.; Anderson, E. A. The Renaissance of Natural Products as Drug Candidates. *Science* **2005,** *310* (5747), 451–453.

Pedersen, M. M.; Chukwujekwu, J. C.; Lategan, C. A.; Van Staden, J.; Smith, P. J.; Staerk, D. Antimalarial Sesquiterpene Lactones from *Distephanus angulifolius*. *Phytochemistry* **2009,** *70* (5), 601–607.

Petrovska, B. B. Historical Review of Medicinal Plants' Usage. *Pharmacog. Rev.* **2012,** *6* (11), 1.

Rao, K. V.; Donia, M. S.; Peng, J., Garcia-Palomero, E.; Alonso, D.; Martinez, A.; Medina, M.; Franzblau, S. G.; Tekwani, B. L.; Khan, S. I.; Wayhuono, S.; Willett, K. L.; Hamann, M. T. Manzamine B and E and Ircinal A Related Alkaloids from an Indonesian *Acanthostrongylophora* Sponge and Their Activity against Infectious, Tropical Parasitic, and Alzheimer's Diseases. *J. Nat. Prod.* **2006,** *69*, 1034–1040.

Rustaiyan, A.; Faridchehr, A.; Bakhtiyari, M. Artimisinin A Sesquiterpene Lactone with an Endoperoxide Bridge Is Toxic to Malarial Parasites. *EJPMR* **2016,** *3* (12), 48–54.

Saxena, S.; Pant, N.; Jain, D. C.; Bhakuni, R. S. Antimalarial Agents from Plant Sources. *Curr. Sci.* **2003,** 1314–1329.

Semple, S. J.; Reynolds, G. D.; O'leary, M. C.; Flower, R. L. Screening of Australian Medicinal Plants for Antiviral Activity. *J. Ethnopharmacol.* **1998,** *60* (2), 163–172.

Shandilya, A.; Chacko, S.; Jayaram, B.; Ghosh, I. A Plausible Mechanism for the Antimalarial Activity of Artemisinin: A Computational Approach. *Sci. Rep.* **2013,** *29* (3), 2513.

Sukumar, K.; Perich, M. J.; Boobar, L. R. Botanical Derivatives in Mosquito Control: A Review. *J. Am. Mosq. Cont. Assoc.* **1991,** *7* (2), 210–237.

Sundaravadivelan, C.; Padmanabhan, M. N.; Sivaprasath, P.; Kishmu, L. Biosynthesized Silver Nanoparticles from *Pedilanthus tithymaloides* Leaf Extract with Anti-developmental Activity against Larval Instars of *Aedes aegypti* L. (Diptera; Culicidae). *Parasitol. Res.* **2013,** *112* (1), 303–311.

Tang, K. F.; Ooi, E. E. Diagnosis of Dengue: An Update. *Expert Rev. Anti-Infect. Ther.* **2012,** *10* (8), 895–907.

Tasdemir, D.; Topaloglu, B.; Perozzo, R.; Brun, R.; O'Neill, R.; Carballeira, N. M.; Zhang, X.; Tonge, P. J.; Linden, A.; Rueedi, P. Marine Natural Products from the Turkish Sponge *Agelas oroides* that Inhibit the Enoyl Reductases from *Plasmodium falciparum*, *Mycobacterium tuberculosis* and *Escherichia coli*. *Bioorg. Med. Chem.* **2007,** *15*, 6834–6845.

Toplak Galle, K. *Domestic Medicinal Plants*; Mozaic Book: Zagreb, 2005; pp 60–61.

Trape, J. F.; Pison, G.; Spiegel, A.; Enel, C.; Rogier, C. Combating Malaria in Africa. *Trend Parasitol.* **2002,** ; *18* (5), 224–230.

Tse, E. G.; Korsik, M.; Todd, M. H. The Past, Present and Future of Anti-malarial Medicines. *Mal. J.* **2019,** *18* (1), 93.

Tucakov, J. *Pharmacognosy*; Institute for Textbook Issuing in SR, Srbije: Beograd, 1964; pp 11–30.

Tucakov, J. Healing with Plants. *Beograd: Rad.* **1990,** 576–578.

Wermuth, C. G.; Villoutreix, B.; Grisoni, S.; Olivier, A.; Rocher, J. P. Strategies in the Search for New Lead Compounds or Original Working Hypotheses. In *The Practice of Medicinal Chemistry*, fourth ed.; Elsevier: Amsterdam, 2015; pp 73–99.

White, N. J.; Nosten, F.; Looareesuwan, S.; Watkins, W. M.; Marsh, K.; Snow, R. W.; Kokwaro, G.; Ouma, J.; Hien, T. T.; Molyneux, M. E.; Taylor, T. E. Averting a Malaria Disaster. *Lancet* **1999,** *353* (9168), 1965–1967.

White, R. D.; Keaney, G. F.; Slown, C. D.; Wood, J. L. Total Synthesis of (±)-Kalihinol C. *Org. Lett.* **2004,** *6* (7), 1123–1126.

WHO (World Health Organization). *Global Strategic Framework for Integrated Vector Management*; World Health Organization: Geneva, 2004a.

Wiart, C. *Etnopharmacology of Medicinal Plants*; Humana Press: New Jersey, 2006; pp 1–50.

World Health Organization. Special Programme for Research, Training in Tropical Diseases, World Health Organization. Department of Control of Neglected Tropical Diseases, World Health Organization. *Epidemic, Pandemic Alert. Dengue: Guidelines for Diagnosis, Treatment, Prevention and Control*. World Health Organization, 2009.

Wright, A. D.; Lang-Unnasch, N. Diterpene Formamides from the Tropical Marine Sponge *Cymbastela hooperi* and Their Antimalarial Activity *In Vitro*. *J. Nat. Prod.* **2009,** *72*, 492–495.

Wright, C. W. Traditional Antimalarials and the Development of Novel Antimalarial Drugs. *J. Ethnopharmacol.* **2005,** *100* (1–2), 67–71.

Wright, A. D.; Wang, H.; Gurrath, M.; Koenig, G. M.; Kocak, G.; Neumann, G.; Loria, P.; Foley, M.; Tilley, L. Inhibition of Heme Detoxification Processes Underlies the Antimalarial Activity of Terpene Isonitrile Compounds from Marine Sponges. *J. Med. Chem.* **2001,** *44*, 873–885.

Wypych, J. C.; Nguyen, T. M.; Nuhant, P.; Benechie, M.; Marazano, C. Further Insight from Model Experiments into a Possible Scenario Concerning the Origin of Manzamine Alkaloids. *Angew. Chem. Int. Ed.* **2008,** *47*, 5418–5421.

Zelefack, F.; Guilet, D.; Fabre, N.; Bayet, C.; Chevalley, S.; Ngouela, S.; Lenta, B. N.; Valentin, A.; Tsamo, E.; Dijoux-Franca, M. G. Cytotoxic and Antiplasmodial Xanthones from *Pentadesma butyracea*. *J. Nat. Prod.* **2009,** *72*, 954–957.

Ziegler, H. L.; Hansen, H. S.; Staerk, D.; Christensen, S. B.; Hägerstrand, H.; Jaroszewski, J. W. The Antiparasitic Compound Licochalcone A Is a Potent Echinocytogenic Agent That Modifies the Erythrocyte Membrane in the Concentration Range Where Antiplasmodial Activity Is Observed. *Antimicrob. Agents Chemother.* **2004,** *48*, 4067–4071.

CHAPTER 2

Targeted Drug Delivery

PRINCY CHOUDHARY and SANGEETA SINGH*

Indian Institute of Information Technology Allahabad, Devghat, Jhalwa, Allahabad 211015, Uttar Pradesh, India

**Corresponding author. E-mail: sangeeta@iiita.ac.in*

ABSTRACT

Drugs are chemical substances that are used to diagnose, prevent, cure, and treat a disease. They are administered in body by various anatomical routes. The conventional methods which are being used since ages are need to be advanced to meet the challenges of new era of diseases. The various disadvantages of ordinary methods like pills, tablets, and ordinary injections can be overcome by targeted drug delivery approach. Targeted drug delivery focuses on delivering the drug or drug–carrier complex at higher concentration to targeted site and at minimum possible concentration to nontargeted sites so as to minimize the adverse effects and maximizing the therapeutic effects. Administration of drugs to specific cavities like pleural cavity, peritoneal cavity, cerebral ventricles, or tissues such as tumors or Kupffer cells of liver, and intracellular localization of drugs or drug carrier system or targeting of DNA and proteins to a cell, all are possible via targeted drug delivery. Carriers which may be polymeric, nanosized particles, lipid molecules, antibodies, or natural cells like erythrocytes or platelets are the main component of targeted drug delivery system which carries drugs and therapeutic moieties to the target sites. Wide range of carriers are available which are specifically used according to the requirements and nature of route and the target site. No doubt, targeted drug delivery has numerous benefits but its clinical applications and feasibility are still under trial. To bring the optimized concepts from labs to clinic, a lot more studies are still needed.

2.1 INTRODUCTION

Drug can be administered in the body by various anatomical routes. For attaining desired therapeutic effects, the importance of choosing most suitable route is undoubted. Hence, there are several factors that need to be considered while planning to deliver a drug-like property of the drug itself, the disease to be dealt with, and the therapeutic time involved. Drugs can be administered via various systemic routes (Fig. 2.1) or can be delivered directly at the targeted site (Coelho et al., 2010).

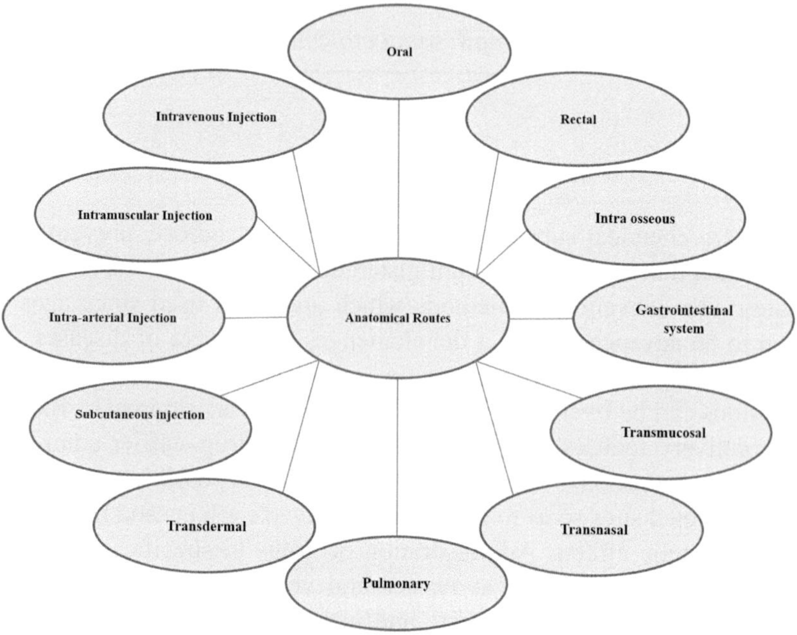

FIGURE 2.1 Various systemic drug delivery routes.

2.2 DRUG ADMINISTRATION: CONVENTIONAL CONCEPT

Drug administration to patients via various routes in the form of pills, capsules, tablets, ointments, cream, injections, liquid, aerosols, and suppositories has been used since decades, but these conventional methods found to be ineffective in maintaining concentration of a drug at the required site. To make this effective, these conventional ways should be applied many times a day that may result in impulsive degradation of drug, drug toxicity, fluctuating level of drug, and ineffective drug concentration (Agnihotri et al., 2011). The scheme drawing of the effect in drug concentration in the body on using different methods of drug administration is shown in Figure 2.2. In 1900, the idea of drug targeting was proposed by a German Nobel Laureate Paul Ehrlich via the concept of magic bullet which he gave (Que et al., 2010). The concept comprises coordination of three fronts: the specific target of the diseased state, a drug that treats the disease effectively, and a carrier which can carry the drug to specific site (Fahmy et al., 2005).

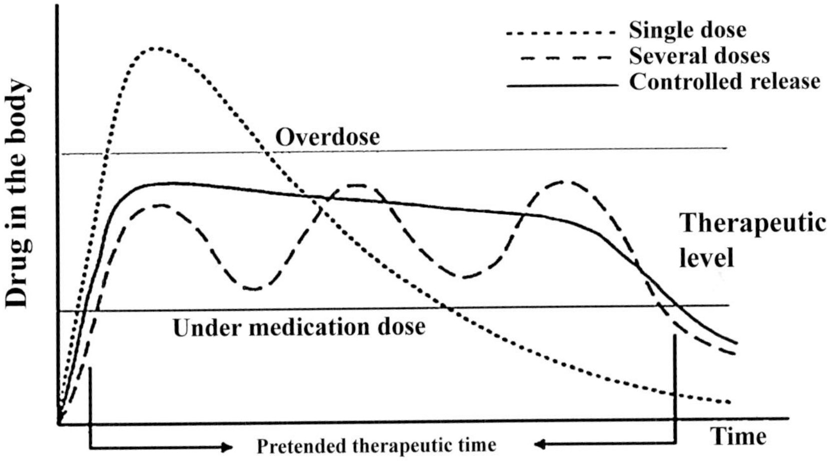

FIGURE 2.2 Scheme drawing of the effect in drug concentration in the body on using different methods of drug administration (Bajpai et al., 2008).

2.3 DRUG ADMINISTRATION: ENGINEERED CONCEPT

Targeted drug delivery is a smart way of delivering drug to the patient such that the drug is assured to reach the targeted site. It is an engineered way to deliver drug at higher concentration to a specific site or target tissue/organ relative to others (Honey, 2018). Its prime objective is to conquer unspecific toxic effects of the conventional drug delivering methods and hence increasing the therapeutic efficacy by reducing the quantity of drug required. Figure 2.3 depicts the concept of targeted drug delivery, and Figure 2.4 outlines the need of targeted drug delivery.

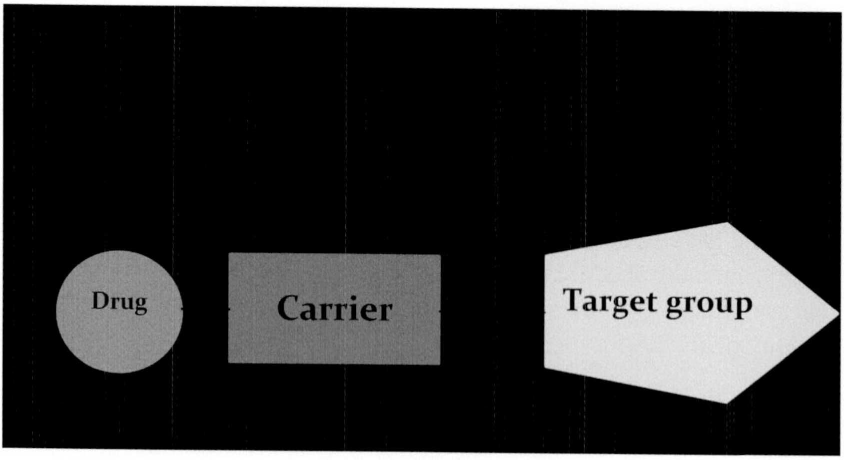

FIGURE 2.3 Concept of a targeted drug delivery.

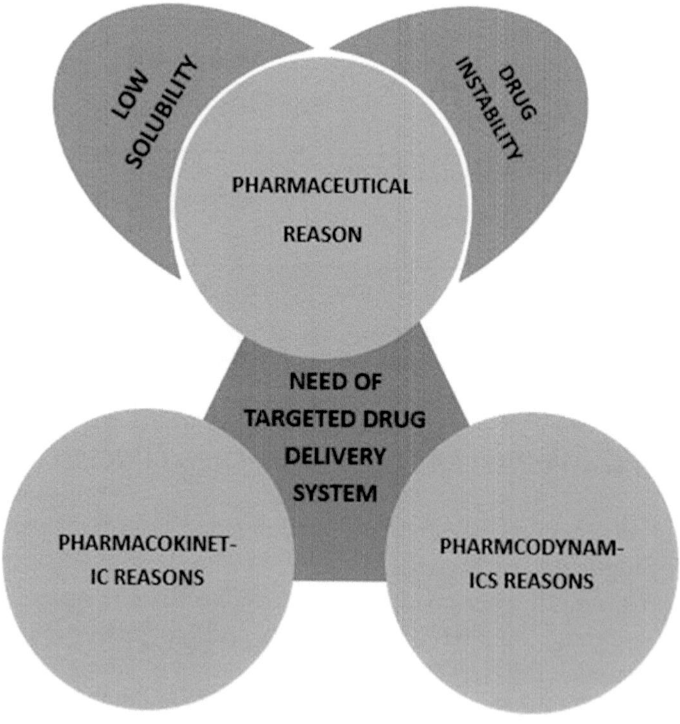

FIGURE 2.4 Need of a targeted drug delivery system.

Targeted drug delivery is different from the conventional methods as in this case; drug gets delivered in the form of dosage while in later case, drug get absorbed in the biological membrane (Mishra et al., 2016). There are four basic requirements for targeted drug delivery system:

i) **Retain:** Proper loading of drug into a suitable carrier that can carry it to the intended site.
ii) **Evade:** Adequate stay in the circulation for reaching the targeted site.
iii) **Target:** Retention by specific feature within targeted site.
iv) **Release:** Release of drug at the meant site at specific time that grant for the effective therapeutic functioning.

Of course, different intended sites in the body require different drug delivery system relating to the route that is selected for delivering drug (Bae and Park, 2011).

2.4 PRINCIPLE AND OBJECTIVE OF TARGETING

The basic idea behind the targeting delivery of drug is to deliver the drug or drug–carrier complex at higher concentration to targeted site and minimum possible concentration to nontargeted sites so as to minimize the adverse effects and maximize the therapeutic effects. The main objectives of drug targeting are depicted in Figure 2.5.

2.4.1 LOCAL AND SYSTEMIC TARGETING

Local targeting is noninvasive in nature, and this approach aims in delivering the drug at the concerned local site for dealing with local pathologies like inflammatory bowel diseases, colon cancer, stomach cancer, gastritis, ulcerative colitis, etc. It primarily confines colon-specific targeted systems, buccal adhesive systems, gastroretentive systems, etc.

In case of systemic targeting, drug is administered via invasive route by intravenous conduction of nanotechnological systems like dendrimers, niosomes, liposomes, liposomes, and SLNs (solid lipid nanoparticle). This system follows systemic circulation for delivering drug. The major constraint of this system is due to drug's pernicious effect to nonspecific tissues (Honey, 2018).

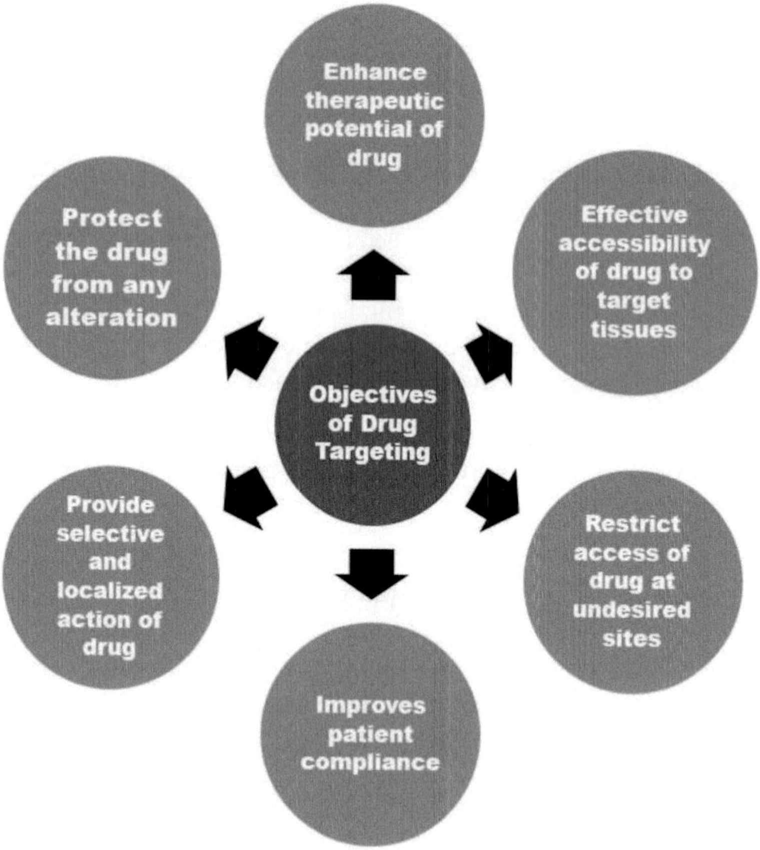

FIGURE 2.5 Objectives of drug targeting.

2.5 TYPES OF DRUG TARGETING

Drug targeting can be broadly classified as shown in Figure 2.6.

2.5.1 ACTIVE TARGETING

It explains particular interactions of drug or drug–carrier with the targeted site. On the basis of particular ligand–receptor interaction that is feasible only when two constituents are in close vicinity (<0.5 nm). Active targeting intends to guide the drug/drug–carrier complex to the target site just like a cruise missile. This "ligand–receptor type interaction" is for intracellular localization that eventuates only after circulation of blood and extravasation. Hence, surface modification of drug is done to increase blood circulation time and boosting EPR effect for enhancing drug delivery to target (e.g., tumor) site (Bae and Park, 2011).

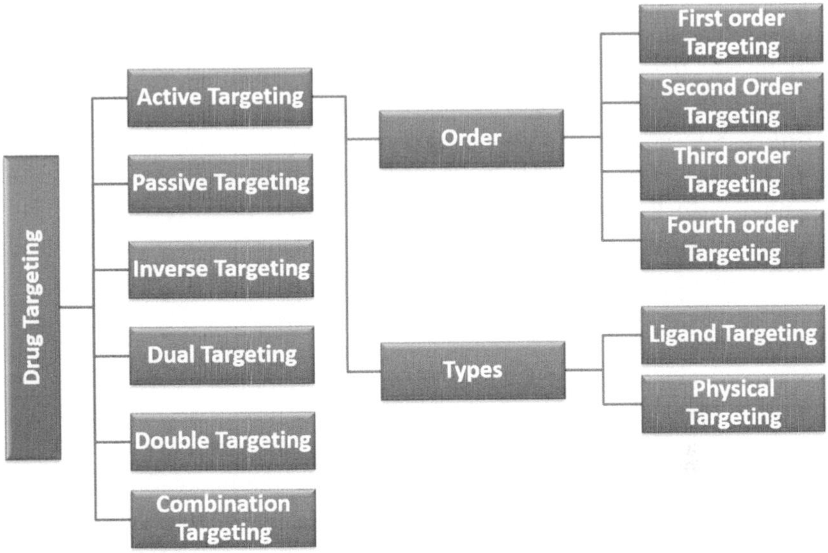

FIGURE 2.6 Classification of drug targeting delivery system.

Active targeting can be divided into four levels:

i) ***First-order targeting:*** It involves dispersion of drugs to the capillary bed of the intended site or organ or tissue as lymphatic tissue, pleural cavity, peritoneal cavity, cerebral ventricles, joints, eyes, etc.
ii) ***Second-order targeting:*** It involves delivering drug to significant tissues like cancerous cells or Kupffer cells in liver.
iii) ***Third-order targeting:*** It involves intracellular localization of drug–carrier complex by endocytosis or via receptor based ligand-mediated approach (Mishra et al., 2016).
iii) ***Fourth-order targeting:*** It involves targeting of drug to DNA or protein like macromolecules (Honey, 2018).

Some scientists have also classified active targeting into ligand targeting and physical targeting.

2.5.1.1 LIGAND TARGETING

The drug–carrier complexes can become active when they get bonded with appropriate molecular ligand that may include antibodies, oligosaccharides, viral proteins, polypeptides, or fusogenic residues. Such engineered systems selectively administer drug to the target tissue/cell. In such targeting, reaction of ligand with its corresponding receptor emphasizes the uptake of whole drug administering system into the target cell/ tissue. Folate receptor targeting is a good example of this approach. Folate receptor is a protein that binds with folic acid with great affinity. Subsequently, receptor binding, folate gets quickly internalized as endosomes via endocytosis. It was noted that the ability of folic acid to bind with the receptor doesn't get affected because of binding of folate with micro- and macromolecules, proteins, radioactive materials, or liposomes; hence uptake of such complexes via receptor mediated endocytosis gets improved.

2.5.1.2 PHYSICAL TARGETING

In this approach, drug carrier is guided toward the targeted site by using physical factors like change in pH, light intensity, temperature, ionic strength, electric field, or glucose concentration. This method was found to be exceptional in case of targeting tumor, delivery of genetic material, or entrapped drugs.

2.5.2 PASSIVE TARGETING

This technique is based on the body's natural ability to react toward the physiochemical properties of the drug or drug–carrier complex. As example, some colloids are easily taken up by the RES (Reticulo Endothelial Systems) in liver and spleen (Agnihotri et al., 2011). It utilizes natural surroundings of the target tissue or site for drug administration as leaky vasculature of tumor tissue helps in accumulation of nanodrugs in tumor tissues. Normally, tumor vessels have numerous pores and are dilated and extremely disorganized that result in widening of gap junctions present between endothelial cells. Within tumor region, the leaky vasculature along with low lymphatic drainage aids in enhancing the permeation and retention of nanoparticles. This is generally referred as EPR effect, i.e., enhanced permeability and retention effect. Preferentially, the drug-loaded nanoparticles get accumulated in tumor tissue than in normal tissues and also as the normal tissues are held up with tight junctions hence, the nanoparticles are not able to cross the capillaries of normal tissues. Therefore, this approach can be effectively used in the accumulation of drug in tumor tissues (Sabyasachi and Sen, 2017). Figure 2.7 clearly explains the differences between active and passive targeting.

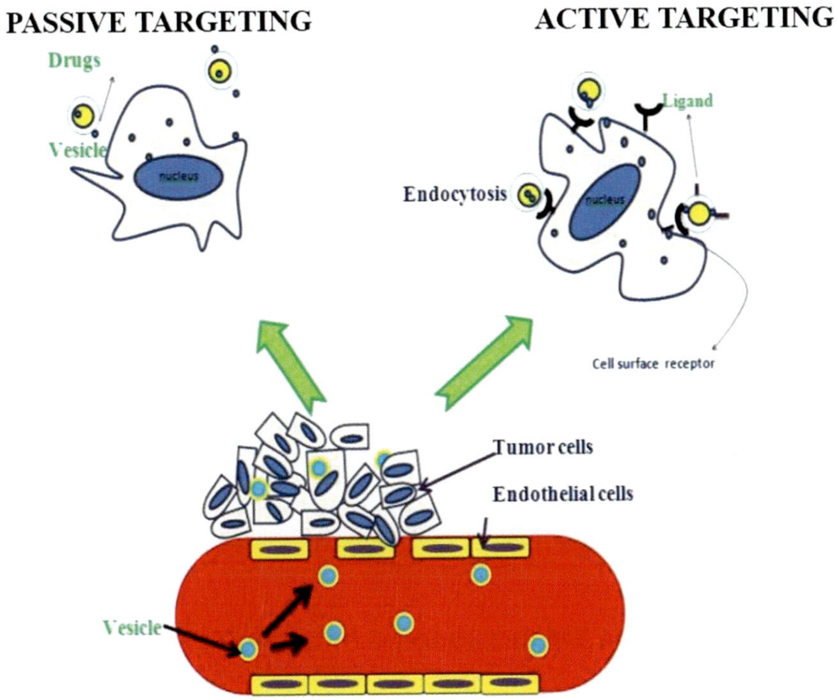

FIGURE 2.7 Active targeting versus passive targeting (Mishra et al., 2016).

2.5.3 INVERSE TARGETING

The approach refers to making attempts to abstain RES from up taking colloidal carrier hence called inverse targeting. For this, normal function of RES is suppressed by preinjection of numerous blank colloidal carriers or dextran sulfate-like macromolecules. This cause RES saturation and suppressed defense mechanism of RES. This is an effective approach for targeting non-RES organs for delivering drugs.

2.5.4 DUAL TARGETING

In this approach, the carrier molecule too has some therapeutic properties which aid on drug properties giving synergistic therapeutic effect. An antiviral drug can be loaded with a carrier having its own antiviral activity which shows synergistic effect.

2.5.5 DOUBLE TARGETING

When spatial and temporal methods gets integrate for targeting a carrier system, the approach is called dual targeting. In spatial delivery, specific organs, tissue or cells, and the subcellular compartments are targeted for drug administration, whereas the rate of delivering the drug is monitored in temporal method (Agnihotri et al., 2011).

2.5.6 COMBINATION TARGETING

In this approach, for directly approaching a target site, the targeting systems are provided with carriers, molecular specific homing devices, and polymers (Kumar, 2017).

2.6 DESIGNING A TARGETED DRUG DELIVERY SYSTEM

There are few characteristics which should be considered while designing a targeted drug delivery system. Like, it should be able to incarcerate drug administration to target site/tissue/organ and should possess a steady capillary distribution. And the polymers which shall be used for carriers should exhibit properties like biocompatibility, chemically inertness, physically robustness as well as should be nonimmunogenic and nontoxic. Drug distribution, release, and drug action should not be affected by the polymeric carriers (Honey, 2018).

2.6.1 COMPONENTS OF TARGETED DRUG DELIVERY

The main components of a targeted drug delivery system constitute a target and carrier for drug or marker. Target is the site (cell/tissue/organ) which is in need of treatment. The path of drug administration requires an important moiety, i.e., drug carrier and subsequently to its leakage from carrier to reach the targeted drug to the specific required site by means of biological metabolism along with its clearance and in addition to not reaching at nontargeted site makes this targeted delivery system more specific with minimized side effects. A special molecule or system which is known as carrier is needed for effectual transportation the drug at targeted site. This is a designed vector that keeps the drug with them via encapsulation or spacer moiety and delivers them into the proximity of a target (Kirti and Paliwal, 2014).

2.6.2 CHARACTERISTICS OF AN IDEAL CARRIER

An ideal carrier should possess certain characteristics which are listed below:

- It should be biocompatible, nonimmunogenic, biodegradable, and nontoxic.
- For nanomedicine, the allowable size range is 10–100 nm. The lower range is determined by the interaction with filtration in kidney and upper size by interaction with RES in liver and spleen. The size of a tumor cell needs to be ~400 nm for accumulating drug.
- There should be least possible leakage of drug before reaching to the required/targeted site.
- The shape of carrier system should be in accordance to get permitted for passage via capillaries.
- At targeted site, there should be controlled release of drug.
- The carrier should get specifically recognized by the targeted cells.
- The carrier should stably hold the drug within.
- It should be designed with ideal biophysiochemical properties for finest drug loading, the circulation half- life in the way, and sustained release of the drug beyond irregular delivery times.
- It should be cost-effective (Honey, 2018).

A carrier can be categorized into different classes as shown below in Figure 2.8.

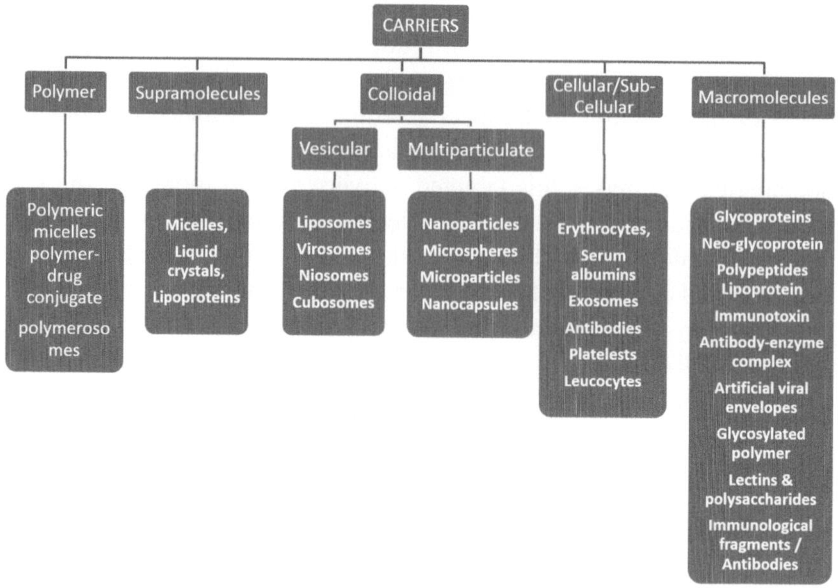

FIGURE 2.8 Carriers used in targeted drug delivery.

2.6.3 *TYPES OF CARRIER*

2.6.3.1 *POLYMERIC CARRIER*

Polymeric micelles, polymer–drug conjugates, and polymerosomes are common carrier in this class. Polymeric micelles are nanosized core/shells like structures formed via amphiphilic block copolymers like Pluromics (polyoxyethylene polyoxypropylene block copolymer self-collaborate to form micelle in aqueous solution). They have a number of advantages like thermodynamics stability as in physiological solution; it leads to slow dissolution *in vivo*. More advantages include size benefits and sustained release (as being in nanometric range they are able to evade RES and can pass via endothelial cells), solubilization of imperfectly soluble molecules (can be an appropriate carrier for water insoluble molecules, such drugs segregate in hydrophobic part and hydrophilic part assists in aqueous dispersion which makes it suitable for intravenous delivery), guarding of encapsulated matter from metabolism or degradation. The most studied block copolymers include poly(ester) s, poly(propylene oxide), and poly (L-amino acid)s (Croy and Kwon, 2006; Srikanth et al., 2012).

Polymer–drug conjugates are the conjugates of the water soluble polymers which are chemically linked with drug via a biodegradable coupler. Drugs are conjugated chemically in such entities which make it different from liposome-like drug vehicles in which drugs are bounded physically. A group of conjugates have been synthesized by implementing linear copolymers in which N-(2-hydroxypropyl) methacrylamide (HPMA) copolymers and polyethylene glycol (PEG) are most studied ones. PEG–protein conjugates are one of the remarkable one as PEG can provide resistance to proteins against enzymatic degradation, and it can also minimize the absorption by reticuloendothelial system. Various FDA-approved drugs like Adagen® (PEG-adenosine deaminase), Oncaspar® (PEG-asparaginase), Pegasys® (PEG-interferon α-2a), PEG-Intron® (PEGinterferon α-2b), and Neulasta® (PEG-granulocyte colony-stimulating factor) are the results of PEGylation of proteins.

Polymerosomes: These are structurally similar to that of liposomes with the exception of the presence of synthetic polymer amphiphiles in them like poly (lactic acid)- or PLA-based copolymers (Mishra et al., 2016; Koren and Vladimir, 2011). Figure 2.9 shows the structures of different polymeric carriers.

2.6.3.2 LIPOSOMES

These are microscopic phospholipid vesicles composed of concentric single or multi-lipid bilayers divided by compartments composed of aqueous buffer in the range of diameter 10–25 μm (Fig. 2.10). They can be SUV (Small Unilamellar Vesicles) (10–100 nm), LUV (Large Unilamellar Vesicles) (100–300 nm) in which single bilayer is present or MLV (multilamellar vesicles) having multiple bilayers. Liposomes are able to deliver drug into cells or even inside subcellular compartments. These are considered to have wider applications in drug delivery as they can deliver from almost every type of route and can be used as carrier for both hydrophilic and lipophilic drugs (Koren and Vladimir, 2011).

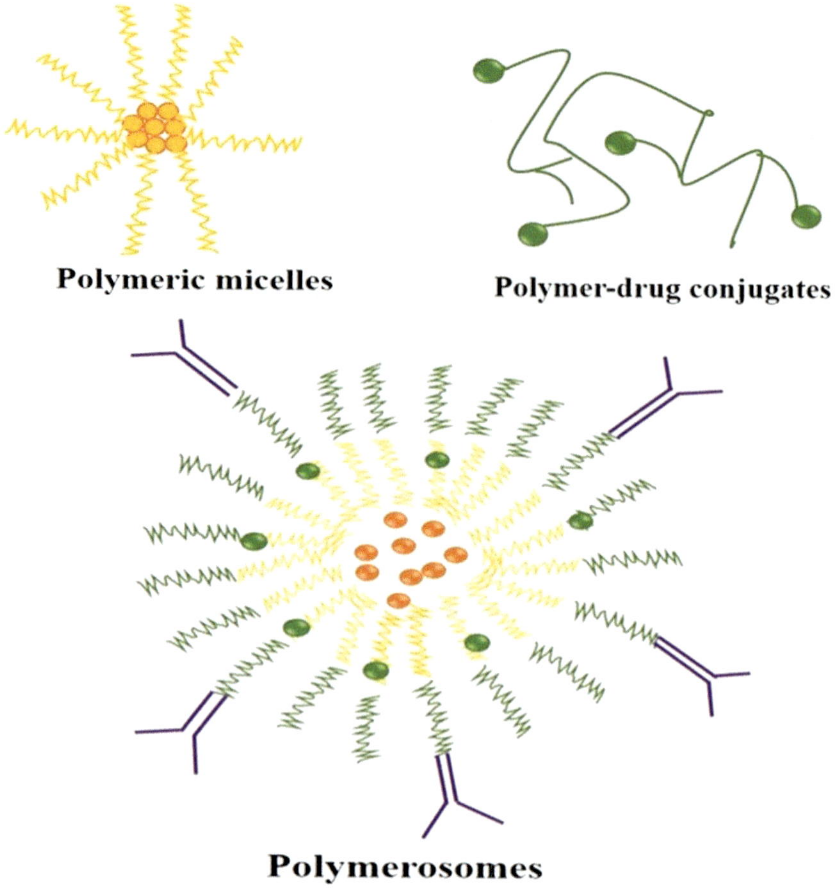

FIGURE 2.9 Structure of different polymeric carriers.

2.6.3.3 NIOSOMES

These are evolved in order to overcome the demerits of liposomes as rancidity, storage, and handling. In this, nonionic surfactants are used instead of lipids which provide them more stability and aid the easiness of handling and storage. Enhancement in physical stability of Niosomes leads to the development of proniosomes which are available in powder state and need to be hydrated with aqueous media and shaken prior to administration (Fig. 2.11).

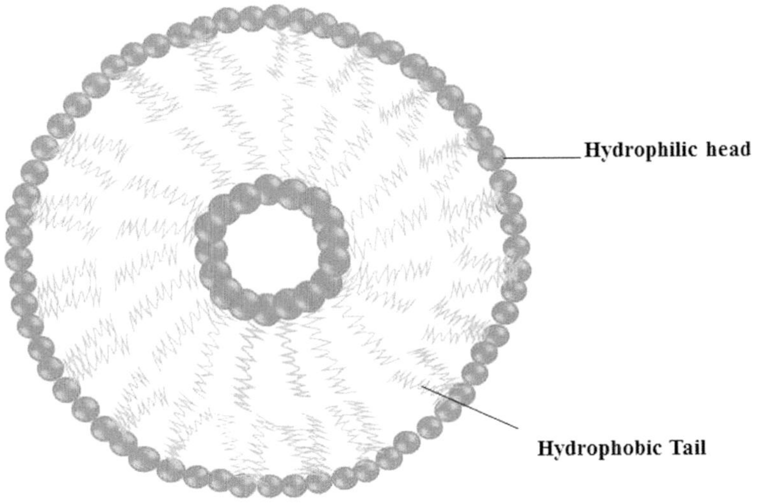

FIGURE 2.10 Structure of a liposome.

2.6.3.4 MICROSPHERE

These are in the free flowing powdered form which is composed of biodegradable synthetic polymers or proteins having particle size <200 nm. These are synthesized for applications like vaccine delivery, controlled drug delivery, or drug carrier by a variety of techniques like polymerization, emulsion, solvent evaporation, spray drying, and spray congealing. These are known for prolonged release of drugs as well as for anticancer drugs targeting to tumor tissues.

2.6.3.5 NANOPARTICLES

Solid particles or particulate matter in the range of 10–1000 nm are categorized under nanoparticles. The drug gets entrapped, encapsulated, or dissolved to get attached with the matrix of the nanoparticle. Nanoparticles have been synthesized mostly by these three approaches: by dispersing preformed polymers, monomeric polymerization, and ionic gelation/by coacervating hydrophilic polymers. They are able to circulate for prolonged time in circulation and are helpful in targeting any specific organ. They can be used as carrier for protein, gene, peptides, or DNA (in gene therapy). All of these advantageous factors make it suitable to use as drug delivery devices. Nanocapsules or nanospheres can be obtained depending on the way of preparing nanoparticles.

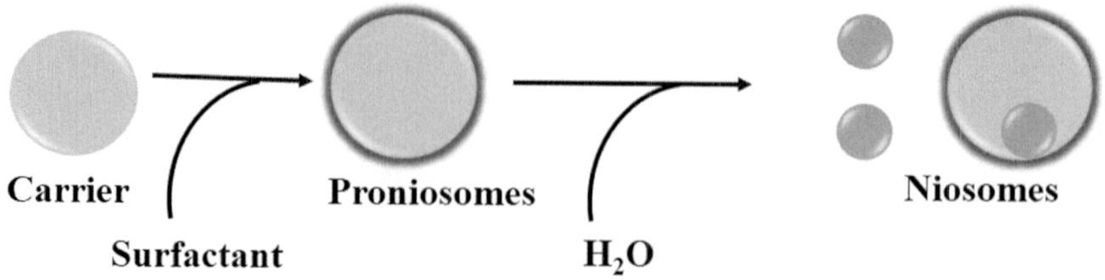

FIGURE 2.11 Formation of niosomes from proniosomes.

Nanocapsules are the structure in which drug is enclosed in a cavity which is surrounded by a polymeric membrane. And nanospheres are a kind of matrix arrangement in which drug is dispersed uniformly and physically (Srikanth et al., 2012).

2.6.3.6 ERYTHROCYTES

Loaded erythrocytes for carrying drug or other therapeutics have been immensely exploited for delivering a wide range of drugs or therapeutic agents by both spatially and temporally controlled methods due to their biocompatibility, biodegradability, and other advantages. As patient's own erythrocytes can be used hence there isn't any risk of any adverse reaction due to foreign negative charges owing mainly to sialic acid's hydroxyl group. The phospholipid present in membrane is nearly 50% of the total content of lipid. The membrane of erythrocyte mainly contains hemoglobin and cytoplasm. Many of the hemoglobin get lost and other components of cell are retained that is termed as resealed erythrocyte which has loosed many of the properties of a normal RBC on resealing. Erythrocyte can entrap an immense type of biologically active substances having size range in 500–600,000 Daltons (Agnihotri et al., 2011; Srikanth et al., 2012).

2.6.3.7 PLATELETS

These are elliptical and nonnucleated disc-like cells which originate in the bone marrow by fragmentation of megakaryocytes. They are being used in treatment of many hematological ailments by carrying drugs and other therapeutic substances. Many drugs like vinka alkaloids, angiotensin, imipramine, and hydrocortisone can bind to platelets (Srikanth et al., 2012).

2.6.3.8 ANTIBODIES

They are being used as drug carriers, target moieties for other carriers, and even as drug itself (Petrak, 2005). Most of the approaches based on recognition of antigen by antibody have been specifically developed for cancer therapeutics. These approaches mainly focused on antigens associated with or expressed by tumor cells. Monoclonal antibodies are also coupled with radioactive markers to detect and visualize the extent of tumor (Agnihotri et al., 2011).

Antibody–drug conjugate is the term given to the conjugate of a drug and a monoclonal antibody that targets selectively for tumor cells or lymphomas. Enzymatic degradation of the linker causes the release of drug under physiological conditions. Mylotarg is an FDA-approved antibody–drug conjugate (Kirti and Paliwal, 2014).

2.6.3.9 LIPOPROTEIN

It is a biochemical arrangement that consists of both lipids and protein (Fig. 2.12). They permit lipids to move via water outside and inside the cell. The function of protein is to emulsify the lipid molecules. Hydrophobic molecules and lipids are transported via plasma lipoproteins in mammalian circulatory system. Because of their smaller size, high residence time, and high drug loading capacity, lipoproteins has the potential to serve as a significant drug delivering vehicle. Lipoproteins include many structural proteins, enzymes, antigens, toxins, and adhesins. These five classes of lipoproteins are found in human plasma: Chylomicrons, VLDL (very low density lipoproteins), HDL (high density lipoproteins), LDL (low density lipoproteins), and free fatty acid-albumin (Srikanth et al., 2012).

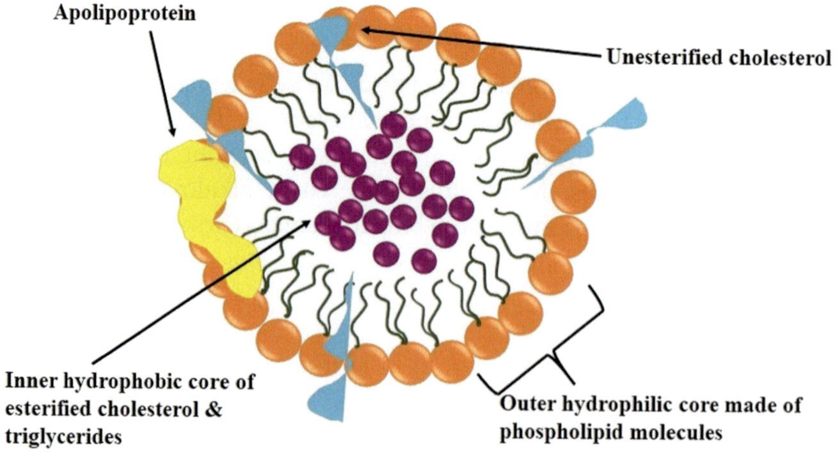

FIGURE 2.12 Structure of a lipoprotein.

2.7 CHALLENGES IN THE DEVELOPMENT OF TARGETED DRUG DELIVERY SYSTEM

- There have not been extensive characterizations of targeted drug delivery system in clinical setting like toxicity, biodistribution, or pharmacokinetics.
- There are challenges regarding uniformity in particle size, product stability, controlling the drug release rate, and manufacturing of sterile preparations on a large scale.
- Optimal features of nanocarriers are determined by screening methodologies which are still not completely understood.
- Because of its unique demands, the development of the targeted delivery system is time consuming and also leads to high cost to benefit ratio.
- The drug–carrier complexes which are administered intravenously face immune reactions.
- The released drug gets diffused and redistributed, and also there are chances of rapid clearance of targeted delivery system by RES clearance system.
- A targeted system gets insufficiently localized into tumor cells.
- The applicability of *in vitro* evaluation of activity, uptake and selectivity, toxicity of the targeted delivery system to animals or human are still under research.

- It is not easy to manufacture ligand targeted systems reproducibly due to their complexity as well as high cost to benefit ratio. It is also quite unrealistic to implement this system in many of the situations.
- The targeted delivery systems are more feasible to liquid cancers comparing to solid ones.
- For bringing the optimized concepts from laboratory to clinic, robust research is need in the complex technology of targeted drug delivery system (Honey, 2018).

KEYWORDS

- **drug**
- **targeted drug delivery approach**
- **tumor**
- **clinical applications**

REFERENCES

Agnihotri, J.; Shubhini S.; Anubha K. Targeting: New Potential Carriers for Targeted Drug Delivery System. *Int. J. Pharmaceut Sci. Rev. Res.* **2011,** *8*(2), 117–123.

Bae, Y. H.; Park, K. Targeted Drug Delivery to Tumors: Myths, Reality and Possibility. *J. Control. Release* **2011,** 153 (3), 198.

Bajpai, A. K., et al. Responsive Polymers in Controlled Drug Delivery. *Prog. Polym. Sci.* **2008,** *33*(1), 1088–1118.

Coelho, J. F.; Ferreira, P. C.; Alves, P.; Cordeiro, R.; Fonseca, A. C.; Góis, J. R.; Gil, M. H. Drug Delivery Systems: Advanced Technologies Potentially Applicable in Personalized Treatments. *EPMA J.* **2010,** *1*(1), 164–209.

Croy, S. R.; Kwon, G. S. Polymeric Micelles for Drug Delivery. *Curr. Pharmaceut. Des.* **2006,** *12*(36), 4669–4684.

Erez, K.; Torchilin, V. P. Drug Carriers for Vascular Drug Delivery. *IUBMB Life* **2011,** *63*(8), 586–595.

Fahmy, Tarek M., et al. Targeted for Drug Delivery. *Mater. Today* **2005,** *8*(8), 18–26.

Honey. Novel Drug Delivery Systems-II. *Pathshala: Pharmaceut. Sci.* **2018,** Paper 08 (Module 26).

Karel, P. Essential Properties of Drug-Targeting Delivery Systems. *Drug Discov. Today* **2005,** *10*(23-24), 1667–1673.

Kirti, R.; Paliwal, S. A Review on Targeted Drug Delivery: Its Entire Focus on Advanced Therapeutics and Diagnostics. *Sch. J. App. Med. Sci.* **2014,** *2* (1C), 328–331.

Koren E.; Torchilin V. P. Drug carriers for vascular drug delivery. *IUBMB life.* **2011**, *63*(8): 586–595.

Kumar, A.; Nautiyal, U.; Kaur, C.; Goel, V.; Piarchand, N. Targeted Drug Delivery System: Current and Current and Novel Approach. *Int. J. Pharm. Med. Res*. **2017**, *5*(2), 448–454.

Mishra, N.; Pant, P.; Porwal, A.; Jaiswal, J.; Samad, M. A.; Tiwari, S. Targeted Drug Delivery: A Review *Am. J. PharmTech Res.* **2016,** 6(1), 2249–3387.

Que Ru. Paul Ehrlich (1854–1915): Man with the Magic Bullet. *Singapore Med J.* **2010**, 51:1–1.

Sabyasachi, M.; Sen, K. K. Drug Delivery Concepts. In *Advanced Technology for Delivering Therapeutics*; InTechOpen, 2017.

Srikanth, K.; Gupta, V. R. M.; Manvi, S. R.; Devenna, N. Particulate Carrier Systems: A Review. *Int. Res. J. Pharm.* **2012,** *3*(11), 22–26.

Ubli, Q. R. Paul Ehrlich (1854-1915): Man with the Magic Bullet. *Singapore Med. J.* **2010,** 1-1.

You Han, B.; Park, K. Targeted Drug Delivery to Tumors: Myths, Reality and Possibility. *J. Control. Release* **2011**, *153*(3), 198.

CHAPTER 3

Targeted Delivery of Biopharmaceuticals for Neurodegenerative Disorders

SARIKA WAIRKAR[1] and VANDANA PATRAVALE[2*]

[1]*Shobhaben Pratapbhai Patel School of Pharmacy and Technology Management, SVKM's NMIMS, V. L. Mehta Road, Vile Parle (W), Mumbai 400056, Maharashtra, India*

[2]*Department of Pharmaceutical Sciences and Technology, Institute of Chemical Technology, Matunga (E), Mumbai 400019, Maharashtra, India*

[*]*Corresponding author. E-mail: vb.patravale@ictmumbai.edu.in; vbp_muict@yahoo.co.in*

ABSTRACT

In the modern era, prevalence of neurodegenerative diseases has considerably grown over the years. Alzheimer's disease, Parkinson's disease, and Huntington's disease are amongst the major neurodegerative disorders that have emerged as a global concern. Therapies for these disorders are undergoing remarkable developments with increasing scope and demand for the same. Various biotechnological products are being explored for the treatment of these neurodegerative disorders. However, physiological barriers like the blood–brain barrier and few critical physicochemical properties of biopharmaceuticals pose a major hindrance to central nervous system drug delivery. Thus, alternative routes of drug targeting like intranasal and transdermal, which also improve the patient's compliance, have been effectively used for these drugs. Similarly, invasive techniques have been successfully applied for brain delivery of biopharmaceuticals. In addition, several colloidal carriers have been explored for passive targeting of biopharmaceuticals whereas more precision has been achieved by active targeting with ligands. This chapter summarizes the advanced delivery approaches of biopharmaceuticals to the brain and the preclinical studies associated with them in treating complex neurodegerative disorders. Nevertheless, a systematic clinical investigation is necessary before exploring their therapeutic translation.

3.1 INTRODUCTION

3.1.1 BIOTECHNOLOGICAL PRODUCTS

An exemplary revolution has been observed in the drug-discovery area in the last few decades and focus of new drug development was shifted from conventional small molecules to biotechnology-based products, also referred to as biopharmaceuticals (Bingham and Ekins, 2009). Although modern biotechnology work has been initiated in terms of recombinant DNA technology in the mid-1970s, the first approved genetically engineered pharmaceutical product by Eli Lilly Corporation, insulin, entered the market in 1982 (Quianzon and Cheikh, 2012). Recently, protein and peptide-based drugs have also caught interest in therapeutics area. Currently, more than 300 biopharmaceutical products are under clinical trials aiming for the treatment of more than 200 diseases. Considering the tremendous growth of this sector, approximately US$314.7 billion global revenue is estimated from the biotechnology industry by 2021 (https://bbsrc.ukri.org/documents/bioscience-facts-figures-pdf/). There is a wide range of biotechnological products including growth factors, cytokines, monoclonal antibodies, enzymes, therapeutic hormones, vaccines, clotting factors, nucleic acid and cell-based therapeutics, gene-based products, and so on. It covers several therapeutic areas like cancer, autoimmune diseases, diabetes mellitus, AIDS, cardiovascular, respiratory, and neurological complications. Additionally, these products are manufactured by an ecofriendly, economical, and speedy production processes (Tang and Zhao, 2009). Commercialization of biotechnological products requires keen knowledge and understanding of the required process conditions. Various bioprocesses including fermentation, enzymatic biotransformation, bioconversion, and so on are implemented for production purposes. From the pharmaceutical product point of view, secondary metabolites obtained from microorganisms using DNA-based technology form a major portion of biopharmaceuticals (Kinch, 2015). In the last few decades, marine species have been widely explored for compounds with therapeutic benefits. With biotechnological products, purity remains a huge challenge and hence advancements in the purification techniques, separation techniques, and so on are being achieved. For successful large-scale production of any biopharmaceutical, molecular engineering techniques in combination with industrial microbiology shall be applied appropriately. Mainly all biotechnological products possess poor bioavailability issues along with severe side effects. Their poor solubility, short half-life, and poor chemical stability make them easily susceptible to degradation on administration and thus targeted delivery of these molecules become a necessity (Rathore and Rajan, 2008). Biopharmaceuticals produced via DNA technology are considered safe and cost-efficient as compared to conventionally produced compounds.

For neurodegenerative diseases, conventional therapy remains inadequate to provide complete cure and therefore, biopharmaceuticals have drawn huge attention for the treatment of these disorders as a next-generation therapy (Albarran et al., 2011).

3.1.2 NEURODEGENERATIVE DISEASES

Neurons are the fundamental units of central nervous systems (brain and spinal cord) and peripheral nervous system (autonomic nervous system and the somatic nervous system). Neurodegeneration is a progressive pathological condition indicating loss of structure or function of neurons and often resulting in neuronal death. Interestingly, neurons cannot regenerate themselves naturally if damaged, attributing to their amitotic nature. This has been especially observed in central nervous system and thus causes permanent loss in the body (Jellinger, 2010). Neurodegenerative diseases are a diverse group of neurological ailments varying in their pathophysiological features, affecting particular subsets of neurons in specific anatomical regions. The common clinical features of these diseases comprise memory and cognitive impairments, speech and breathing difficulties, and motor dysfunction. Common neurodegenerative disorders include Alzheimer's disease (AD), PD, dementia with Lewy bodies, Huntington's disease (HD), progressive supranuclear palsy, amyotrophic lateral sclerosis (ALS), multiple system atrophy, corticobasal degeneration, and so on (Gitler et al., 2017). Few common and more prevalent neurodegenerative diseases are described in the following section.

3.1.2.1 ALZHEIMER'S DISEASE

AD is the most common neurodegenerative disease, clinically characterized by progressive cognitive dysfunction and dementia. Neuroinflammation accompanying with the deposition of amyloid-beta (Aβ) in brain is an important feature of AD pathology. Increased expression of cytokines, reactive oxygen species, microglial activation, and nuclear factor kappa B (NF-kB) are the major contributors of inflammatory process of AD (Kumar et al., 2015). It is estimated that the global incidence of dementia is as high as 36 million and will reach 66 million by 2030 and 114 million by 2050, with approximately two-thirds of those patients living in developing countries (Alzheimer, 2015).

Due to complexity of AD, its etiology and pathophysiology are still not clear. Thus, current treatments of AD give only symptomatic relief rather than complete cure. Donepezil, rivastigmine, and galantamine are most widely used cholinesterase inhibitors for Alzheimer's treatment along with NMDA antagonist, memantine, in a few cases (Tan et al., 2014). Besides, therapeutic concentration of the drug is constrained due to blood-brain barrier (BBB) and low brain permeability of drugs (Weiss et al., 2009). In this context, nose-to-brain pathway has been the topic of interest for formulation scientists for effective drug delivery to the brain. Bypassing the BBB and targeting brain via olfactory and trigeminal neural pathways to enhance efficacy of neurotherapeutics have been extensively reported in the literature (Pardeshi and Belgamwar, 2013).

3.1.2.2 PARKINSON'S DISEASE

PD is a progressive degenerative disorder of central nervous system (CNS) commonly observed in aged patients. The prevalence of PD is more than 200 patients per 100,000 with maximum cases of patients more than 50 years of age. A hallmark feature of PD is neurodegeneration particularly in substantia nigra pars compacta and the nigrostriatal tract of brain.

The degeneration of the terminals of dopamine (DA) neurons results in decreased affinity for uptake of DA. Consequently, there is deficiency of DA in the striatum which controls muscle tone and coordinates movements. Therefore, there is an imbalance in dopaminergic and cholinergic system in striatum that leads to motor defects. Though several groups of dopaminergic neurons are present in CNS, the loss of DA cells in the substantia nigra pars compacta is believed to be the reason for all motor manifestations of PD (Imamura et al., 2006). The specific cause of degeneration of nigrostriatal neurons is unknown, but believed to be multifunctional. It is characterized with four cardinal motor manifestations; tremors at rest, muscle rigidity, bradykinesia, and postural instability (Maiti et al., 2017).

Current therapy for PD is essentially symptomatic. Several therapeutic agents used in management of PD include levodopa (L-DOPA), selegiline, rasagiline, pramipexole, amantadine, and bromocriptine. A natural DA precursor, L-DOPA, is the gold standard of anti-Parkinson treatment (Wei et al., 2010). However, L-DOPA's efficacy in the latter stages of PD is considerably decreased due to its metabolism, low bioavailability, and subsequently variations in the plasma levels. The major peripheral metabolism of L-DOPA demands administration of large doses of the L-DOPA which causes adverse effects like nausea, vomiting, cardiac arrhythmias, and hypotension. Thus, combination therapy of L-DOPA with extracerebral DOPA decarboxylase inhibitors (e.g., carbidopa and benserazide) has been recommended to reduce reduction in the levodopa dose (Pinder et al., 1976). Similarly, the use of carboxy-*o*-methyl transferase inhibitors (tolcapone and entacapone) has been advised to reduce fluctuations in plasma concentrations of L-DOPA, whereas selegiline is administrated as an MAO (monoamine oxidase) inhibitor (Antonini et al., 2008). Hence, these shortcomings of pharmacotherapy result in unavailability of a satisfactory cure for PD patients.

3.1.2.3 MULTIPLE SCLEROSIS

Multiple sclerosis is a chronic, autoimmune disease with concomitant inflammation, demyelination, and damage to white matter that covers the axons. The etiology of MS is still not very clear but it is believed that a virus or an antigen may trigger the immunological system to attack its own myelin sheath. Environmental factors may contribute to the pathology of

MS along with genetic influences. However, unlike other neurodegenerative disorders, protein aggregates are not reported as contributing factors for MS. The clinical manifestations of MS associated with neurological weakening include motor weakness, visual impairment, ataxia, dysarthria, diplopia, and so on, and cognitive deficits related to memory, attention, concentration, and so on. It is prevalent in young population and more predominant in females than males (Chaudhuri, 2013).

3.1.2.4 HUNTINGTON'S DISEASE

HD is a monogenic disorder that causes progressive degeneration of nerve cells in the brain. The defective gene produces atypical form of Huntington protein which is toxic and leads to breakdown of nerves over a period of time. Being a hereditary disease, there is always a risk of HD development in the progeny irrespective of gender. HD patient shows characteristic symptoms like cognitive deficits (impaired memory and judgment), motor disorders (shaky walk and involuntary movements), and behavioral changes (mood swings, personality changes, and depression) along with trouble in swallowing and slurred speech, breathing problems, weight loss, and so on (Bano et al., 2011).

3.1.2.5 AMYOTROPHIC LATERAL SCLEROSIS

ALS, also known as Lou Gehrig's disease, is a very rare neurodegenerative disease, which affects motor neurons that control the voluntary muscles. It is a progressive disorder wherein the nerve cells gradually weaken, break down, and die later. ALS is inherited by approximately 5–10% of patients; however, the specific causes remain unknown for remaining population. There are multiple possible causes reported for ALS including genetic mutation, chemical imbalance in the brain, disordered immune response, head injury/trauma, and environmental factors such as smoking, exposure to environmental toxins, metal ions, and so on. The early symptoms of ALS include muscle weakness or muscle atrophy, cramps, stiffness of affected muscles, troubled swallowing or breathing, and so on. At a later stage of ALS, symptoms worsen to severe breathing problems that often lead to respiratory failure, frontotemporal dementia, and aspiration pneumonia. Although there is no cure for ALS, the symptomatic treatment and supportive care can be provided to improve survival and quality of patient's life (Van et al., 2017). Figure 3.1 illustrates the summary of microscopic and macroscopic features of AD, PD, MS, HD, and ALS.

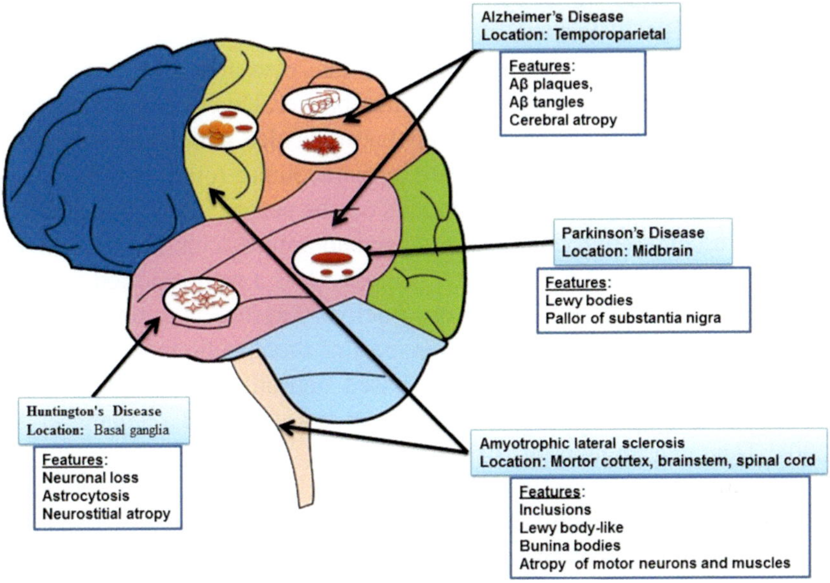

FIGURE 3.1 Illustration of anatomical locations and features of neurodegenerative disorders.

3.2 OBSTACLES IN CNS DELIVERY

The greatest challenge in effective therapy of neurodisorders is penetration of drug into the brain and maintaining therapeutic concentration at target site for a sustained period. This is attributed to distinctive anatomical and physiological features of nervous system and complex architecture of brain. The brain is protected from possibly toxic substances by the presence of two barrier systems, namely the BBB and the blood cerebrospinal fluid barrier. Unfortunately, the same shield which prevents entry of harmful chemicals also constrains therapeutic agents. Moreover, the inadequate regeneration capacity of brain tissue also contributes for limited therapeutic benefits of potential drug candidates.

3.2.1 BIOLOGICAL BARRIERS

3.2.1.1 BLOOD–BRAIN BARRIER

BBB is a semipermeable, physiological barrier that exists around the capillaries of brain which controls the transportation of molecules across the brain. BBB is formed by endothelial cells (ECs) of the capillary wall, astrocyte, and pericytes at multiple neurovascular sites. Especially, astrocytes determine the tightness of this barrier. Due to continuous tight junctions, BBB restricts the entry of polar molecules, pathogens, hazardous substances in the blood, transported by paracellular route. However, few gases, water, and lipophilic compounds can be passed freely though BBB. Besides, essential nutrients like glucose and amino acids can be diffused through it with specific transporter proteins. It is a well-reported fact that the major barrier of an effective drug-delivery system targeting the brain is BBB. It hinders almost 100% transport of therapeutic macromolecules and above 98% of small molecule-based neurotherapeutics. Therefore, numerous attempts have been made to overcome BBB (Ballabh et al., 2004; Tietz and Engelhardt, 2015).

There are a variety of changes observed in BBB during CNS disorders which may act as a hallmark for the disease. These changes in BBB also alter the quality of obstruction provided by BBB and subsequently lead to development of disease. In case of AD, there is a defect in the transport of amyloid-β across the BBB, and during obesity, development of leptin resistance is observed which leads to reduction in transport of leptin. The toughness of BBB reduces with age which leads to an alteration in pharmacokinetics of the drug, brain penetration, and also enhances the drug–drug interactions. The transport across BBB is largely dependent on physicochemical properties of the drug. The properties which reduce BBB transport of the drug include higher branching in the structure of the drug, high polar surface area, higher tendency to form hydrogen bonds, and low lipid solubility. BBB also acts as an enzymatic barrier which adds onto the challenges in drug delivery since the drugs are directly exposed to cytosolic and membrane-associated enzymes like alkaline phosphatase, aminopeptidase, dipeptidyl(amino)-peptidase IV, aromatic acid decarboxylase, and *c*-glutamyltranspeptidase which metabolize the neuroactive agents in the BBB (Zlokovic, 2008; Carvey et al., 2009).

3.2.1.2 BLOOD CEREBROSPINAL FLUID BARRIER

Each cerebral ventricle contains choroid plexuses, formed by modified ependymal cells. The capillaries of choroid plexuses are not tightly arranged and permit free movement of small molecules, in contrast to BBB. Blood cerebrospinal fluid barrier (BCSFB) is formed by epithelial cells and the tight junctions that link them. Also, the arachnoid membrane is an important part of BCSFB which wraps the brain and has tightly arranged cells. The choroid epithelial boundary of the BCSFB has structural distinguishing features that assist regulatory functions for supporting neuronal wellbeing. BCSFB reduces penetration of drugs that are transported by paracellular route into cerebrospinal fluid. Hence, drugs need to adopt transcellular route of transport. The drug properties affecting ability to cross BCSFB include molecular weight, hydrophilicity, and intrinsic permeation across the lipid bilayer. High molecular weight, high hydrophilicity, and low intrinsic permeation across the lipid bilayer reduce access of drug to BCSF (Kusuhara and Sugiyama, 2001). The permeation across BCSFB increases with age.

3.2.2 PHYSIOLOGY OF BBB AND DRUG TRANSPORT

BBB acts as a barrier for metabolism and transport of drugs. It is capable of responding to signals from both blood and brain. The gaseous nature of carbon dioxide and oxygen and other gases like helium, xenon, nitrogen makes them easily accessible to the BBB. The ability to access the BBB depends on polarity of molecules. Polar compounds are not accessible to BBB whereas the nonpolar molecules are easily accessible to it. For the entry of polar molecules like glucose and amino acids, there is an underlying requirement of specialized transporters. There are variety of efflux transporters present on the luminal membrane of blood capillaries. The ATP-binding cassette (ABC) transporters are responsible for movement of lipophilic compounds out of the brain. They work against the concentration gradient. P-glycoprotein (PgP) efflux pump (ABCB1) and the breast cancer resistance protein (ABCG2) also play a vital role in drug efflux out of the brain, where ABCC1 is important for efflux from the cerebrospinal fluid (CSF) to the blood. Various alterations are seen in BBB in terms of activity of these transporters. In epilepsy, there is an overexpression of PgP efflux pump in BBB and in neurodegenerative diseases like Alzheimer disease and PD there is alteration in the activity of ABC transporters (Loscher and Potschka, 2005; Tietz and Engelhardt, 2015).

The possible mechanisms for transport across BBB (Fig. 3.2) include

1. Passive diffusion
2. Carrier-mediated transport
3. Receptor-mediated endocytosis/transcytosis
4. Adsorptive-mediated transcytosis (AMT)

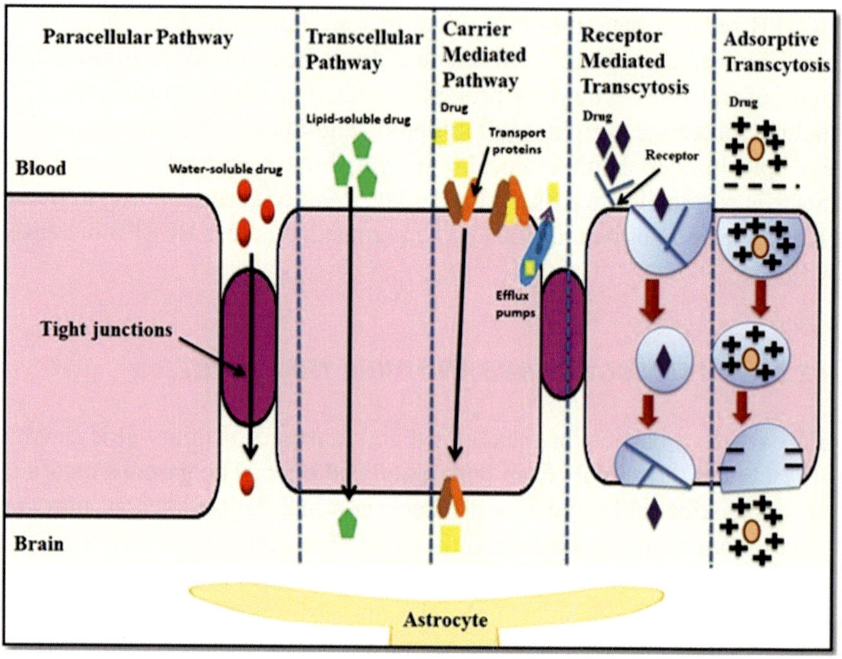

FIGURE 3.2 Mechanisms of transport across BBB.

From the available transport mechanisms, AMT is most effective for the delivery of macromolecules and antibodies, while passive diffusion is of importance in delivering low molecular weight (<300 Da) lipophilic drugs. To deliver polar molecules which do not easily diffuse across the cell membrane, carrier-mediated systems are used. The delivery of nanoparticulate systems is via receptor-mediated endocytosis.

3.2.2.1 CNS DRUG TRANSPORTER AFFINITY

It is well known that 15% of all genes expressed selectively at BBB encrypt for transporter proteins. Presently more than 50% of BBB transporters are not studied scientifically. The separation of CNS and non-CNS therapeutics in WDI (World Development Indicators) has aided in the determination of parameters pertaining to passive BBB permeability and PgP affinity. A major obstacle for the entry of molecules in CNS is PgP and the compound affinity to PgP is determined using PgP efflux ratios. In-vivo rat brain penetration could not correlate with the filters based on PgP efflux ratio from the cell culture permeability assay of Caco-2 colonic cell. More than 1500 CNS and non-CNS drugs were considered to create a model for permeation via passive diffusion and to determine physicochemical criteria for PgP substrates. The use of simple descriptors was adequate to evaluate BBB permeation. There are a variety of in-silico models that are recently reviewed to predict permeation across the BBB (Goyal et al., 2014).

3.2.3 PHYSICOCHEMICAL PROPERTIES OF DRUG

The penetration of CNS drugs is affected by their physicochemical properties and their affinity towards CNS transporters, mainly PgP efflux transporter. Most of the drugs used in treatment of neurodegenerative diseases are typically lipophilic small molecules with molecular weight of <500 Da, so they can pass through BBB. Although few polar drugs or large-molecule drugs are potential candidates for neurotherapy, they cannot be used due to their undesirable physicochemical characteristics. Apart from its size, partition coefficient (log P), is a measure of relative solubility of drug in hydrophilic and hydrophobic solvents. Similarly, drugs which get ionized at the physiological pH of 7.4 would not diffuse in the brain. According to Lipinski's Rule of Five, drug should not possess more than five hydrogen bond donors and more than 10 hydrogen bond acceptors. Neuropharmaceuticals having molecular weight between 150 and 500 Da and a log partition coefficient between 0.5 and 6.0 are frequently used. As far as chemical structure is concerned, drugs with less than five nitrogen and oxygen atoms have a greater possibility of diffusion into the brain (Rankovic, 2015). As reported by Kelder et al. (1999), the value of polar surface area of CNS targeted drugs should be less than 60–70.

3.2.4 PHARMACOKINETIC ASPECTS AND DRUG PERMEABILITY

The extent of absorption in CNS is dependent on the ability of drugs to cross BBB and BCSFB, resists enzymatic degradation and its route of transport. With higher BBB and BCSFB permeability, the drug is more likely to be absorbed. To enhance absorption of drugs acting on CNS, they are combined with peripheral decarboxylase inhibitor which helps in reducing their enzymatic degradation. An example of such drugs is L-DOPA. Route of administration also plays a critical role in deciding the rate of absorption of drugs in CNS. Delivery of drug in the form of an intracranial injection or infusion makes it easier for the drug to circumvent BBB, thereby increasing its absorption rate.

The distribution of drugs in CNS is greater since brain is a highly vascular tissue. The lipophilic nature of drug further enhances its distribution. To determine the extent of distribution, plasma and brain unbound fractions are evaluated. Higher the unbound fractions, lesser is the distribution.

The presence of CYP enzymes in different regions of brain aids in drug metabolism. Highest CYP content is reported in brain stem and cerebellum, whereas striatum and hippocampus show lowest CYP content. In brain cells, CYPs are present in endoplasmic reticulum and inner membrane of mitochondria. CYP isoforms in brain also vary with location. CYP1B1 is found to a higher extent in spinal cord, temporal and frontal cortex, medulla oblongata and putamen, whereas in cerebellum, thalamus, hippocampus, and substantia nigra, it is low. CYP2D6 is mainly present in substantia nigra while lesser amount of it is seen in hippocampus, cerebellum, and putamen. Enzymes like peripheral decarboxylases, monoamine oxidases, catechol-*O*-methyl transferase also aid the metabolism of CNS drugs. Excretion of CNS drugs primarily takes place in liver and kidneys. For example, L-DOPA is primarily excreted in the urine after metabolism and biotransformation (Alavijeh et al., 2005; Nwogu et al., 2016; Ghose et al., 2012).

3.3 BIOPHARMACEUTICALS FOR NEURODEGENERATIVE DISEASES

The delivery of drugs targeting CNS is an interesting area to explore various opportunities and challenges faced to provide better efficacy and desired therapeutic action. Numerous novel agents and delivery routes are formulated and investigated day by day aiming minimal dose and fewer side effects. To achieve it, enhanced permeation, maximized absorption, and retention of drug across several barriers is to be expected and determined. Research scientists are in long-term scrutinization to find ways to get the drug into brain. From planting Trojan horse to using ultrasound to disrupt BBB temporarily are current successful therapy strategies for CNS disorders. The incorporation of advanced biotechnological products to treat neurodegenerative brain disorder appears to be a promising approach in research and development field with various techniques, delivery vehicles, or routes (Mitragotri et al., 2014).

3.3.1 ALTERNATIVE ROUTES OF ADMINISTRATION FOR CNS TARGETING

3.3.1.1 INTRANASAL ROUTE

There are therapeutic agents which cannot reach CNS from systemic blood circulation due to restriction by BBB and also blood–cerebrospinal fluid barrier. To bypass such barriers, olfactory pathway, nose to brain route, can be effective to improve drug delivery.

Various developed protein and peptides meant for treating neurodegenerative disorders are facing obstacles in delivery due to issues like instability, low gastrointestinal (GI) absorption, fast renal elimination, and rapid enzymatic metabolism. To overcome this issue, genetically engineered biotechnological products (interferons, antibodies, etc.) can be utilized in delivering therapeutic drugs into the brain (Henkin, 2011).

To evict accumulated plaque, decelerate neuronal impairment and mend cognitive activity in middle staged patients of AD, a peptide with polyarginines and a unit from main region of Aβ amyloid, that is, R_8-Aβ was developed and investigated on amyloid precursor protein (APP)/ presenilin 1 (PS1) double transgenic mice (Carvey et al., 2009; Henkin, 2011). The results demonstrated not only adjourning the plaque and amyloid formation but also suggested extended use of therapeutic peptides for other neurological and neurodegenerative diseases (Cheng et al., 2017). Intranasal insulin-like growth factor-I (IGF-1) of 7649 Da molecular weight formed of 70 amino acids is a neurotropic factor structurally homologous to proinsulin. It is mainly found in liver and in various other tissues/organs beside CNS. In adult rats, highest amount of IGF-I binding sites are located in olfactory region, circumventricular organs, frontal cortex, cerebellum, and so on. Administration of IGF-I to CNS (ranging between 0.1 and 10 nM) has demonstrated maximum protective and stimulating action corresponding to induction of neuronal survival. This may be attributed to protection of hippocampus region from β-amyloid neurotoxicity, rescue of cortical neurons from nitric oxide-led neurotoxicity, minimized tau phosphorylation along with instigated neurogenesis. An investigation on anesthetized adult SD rats was conducted by delivering IGF-I intranasally and intravenously separately. The results of intranasal administration displayed IGF-I circumventing BBB through olfactory and trigeminal-based extracellular pathway, mustering desired therapeutic actions at multiple regions of brain and spinal cord (Thorne et al., 2004; Avgerinos et al., 2018).

Intranasal route for delivering nerve growth factor (NGF) to the brain seems to be an alternative and innovative remedial approach to seek cure for neurodegenerative disorders (Aloe et al., 2012). Numerous studies on animals and primates have revealed the neuroprotective activity exhibited by NGF against cholinergic system deficit, tau hyperphosphorylation and β-amyloid neurotoxicity (Covaceuszach et al., 2009). The feasibility of NGF can be improved by increasing NGF therapeutic window via averting nociceptive effects. To achieve desirable output, variant form of NGF was mutated with R100 residue directed by Hereditary Sensory Autonomic Neuropathy Type V (HSAN V), human genetic disease which allows maximizing the NGF dose without causing postpain effects. This hNGF mutant arise to be a rationale for noninvasive system when administered to APP/PS1 mice that displayed absolute neurotrophic property, anti-amyloidogenic action, decreased nociceptive effects, thus exhibiting overall progress in aborting neurodegeneration phenomena (Capsoni et al., 2000). Similarly, when NGF was administered to AD11 mice by intranasal route, memory and recognition impairment, behavioral and cognitive deficit and cholinergic inadequacy was observed to be gradually reverting while assessing performance in object recognition and memory test (De Rosa et al., 2005). Neuropeptide, hypocretin-1, also called as orexin-A is produced by neurons of hypothalamus, whose major role is to regulate sleep–wake cycle and appetite physiologically. Intranasal targeting of hypocretin to CNS was conducted on male SD rats and was compared to intravenous delivery system. Though intranasal administration resulted in 10-fold lower hypocretin level in blood concentration than intravenous but was significantly higher in tissue—blood concentration that is, 14-fold and 9-fold greater in trigeminal nerve and olfactory region, respectively. Delivering hypocretin through intranasal route increased drug targeting from 5 to 8 fold (Dhuria et al., 2009). Neurodegenerative disorders cause disabilities which prove to be the major cause of death in elderly patients. Overcoming these disabilities is a major challenge due to existence of BBB in the CNS. Using basic fibroblast growth factor (bFGF), the neuronal loss in hippocampus region could be minimized and learning deficits in AD could also be improved. Radiolabeling of bFGF was done using Na ^{125}I iodogen. Radiolabelled bFGF was incorporated in lectin modified polyethylene glycol-polylactic-*co*-glycolic acid (PEG–PLGA) nanoparticles and coated with *Solanum tuberosum* which binds to *N*-acetylglucosamine present in the nasal epithelium. The method used for preparation was emulsion/solvent evaporation method. The brain distribution studies were done using radiolabeling which indicated that more than 70% of bFGF was transported to brain via nose–brain direct pathway. Intranasal administration of these particles in different AD model was performed to check neuroprotection effects. Morris water maze test showed increase in the spatial learning and memory capability in animals. Choline acetyl transferase activity in rat hippocampus demonstrated that bFGF formulations were effective in recognition and memory improvement. Histology studies also indicated neuroprotection in the hippocampus region of AD rats. In-vivo safety evaluation studies demonstrated that no toxic effects were seen in spleen, liver, heart, and lungs. Intranasal administration reduced toxicity to kidney when compared to intravenous administration. The study showed that nanoparticles coated with *S. tuberosum* can be effective carries for proteins and peptides like bFGF to enter CNS and exhibit their therapeutic action (Zhang et al., 2014). Erythropoietin has been identified as a neuroprotective cytokine or glycoprotein against neurodegenerative conditions (Fernando et al., 2018). When neuro-erythropoietin (EPO) (125, 250 μg/kg) was administered intranasally to 12–14 months aged nontransgenic mouse model of AD thrice a day and thrice a week, the motor, and memory responses were found to be ameliorated as depicted by performance evaluation using water-maze and novel-object recognition test. Also, increase in lipid peroxidation appeared to be diminishing with the results of immunohistochemical and enzyme-linked immunosorbent assay (ELISA), thus illustrating significant declination of Aβ1–42 deposits. This justifies the neuroprotective action of neuro-EPO on intranasal delivery without any unwanted effects (Rodriguez Cruz et al., 2017). Delivery of genes has the biggest limitation over selecting suitable viral vectors as majority of them are too large to be directly injected into brain. Vector-based delivery using gene herpes simplex virus-2 is neurotropic and less virulent. An investigation proved this by protecting rats and mice from seizures and neuronal damage. This forms a therapeutic foundation for neurodegenerative disorders treatment.

On obtaining such accomplishment, several more clinical trials for investigating drugs via intranasal delivery are currently carried out for neurodegenerative disorder therapy.

3.3.1.2 TRANSDERMAL ROUTE

Skin offers primary defense as it has T-cell abundance over all body; thus, vaccines can be best alternative therapy to overcome the progression of fatal neurodegenerative disease (Larregina et al., 2001). Th-2 dominant immune response can be prompted by transcutaneous immunization (TCI) which tends to be a remedy for AD (Nguyen et al., 2017). In contemplation to gain this response, vaccine efficacy in transgenic mouse model (APP/PS1) of Alzheimer disease was investigated using cone-shaped MicroHyalamicroneedle (MH800 and MH300) as TCI device with base material sodium hyaluronate consisting $A\beta_{1\text{–}42 \text{ and } 1\text{–}35}$ and cholera toxin as adjuvant respectively where each immunization started at 4 months of age to the skin of the back. Initially, for the first month, vaccination to APP/PS1 mice was performed on weekly basis, later biweekly for following 12 weeks. Serum was obtained from vaccinated mice at marked points and concentration of anti-$A\beta_{1\text{–}42}$ was determined using ELISA. Morris water maze test, novel object recognition assay, brain section immunostaining, and quantitative estimation of Aβ oligomers/fibrils (i.e., $A\beta_{1\text{–}40}$ or $A\beta_{1\text{–}42}$) in brain were also performed. The results demonstrated effects of TCI group with $A\beta_{1\text{–}35}$ showing minor recovery in cognitive activity and augmented immunogenicity than $A\beta_{1\text{–}42}$. Thus, it is reported that amyloid immunization can also eliminate β-amyloid plaque accumulation even at later stages of disease reducing the progression and magnitude of neuropathology (Schenk et al., 1999). Recent studies also illustrated that both anti-Aβ1–42 IgG and epitopes or isotypes are vital for eliminating senile plaques and fixing cognitive function (Matsuo et al., 2014). Transdermal patches are a promising approach meant for frequently dosed medication to provide community-based healthcare. Delivering therapeutic agents through skin is a noninvasive system bypassing gastrointestinal limitations and first-pass metabolism. The advancement of transdermal drug-delivery system has steered remedial measures for several neurological and psychiatric disorders such as depression, PD, ADHD, and dementia. In virtue of its long-acting property, constant plasma level of drug can be maintained which is requisite against episodic withdrawals (Isaac and Holvey, 2012). Rivastigmine, cholinesterase inhibitor served in AD therapy is available in the form of transdermal patches which caters dementia-led Alzheimer and PD patients and provides an easy and convenient way to administer medicine over other conventional formulations like tablets, capsules. Orally administered rivastigmine formulation displays cholinergic gastrointestinal side effects induced by sharp increase of acetylcholine level. Pharmacokinetic studies of transdermal administration revealed prolonged duration of Rivastigmine to reach plasma peak level which minimizes irregularities in plasma concentration and reduced associated after effects (Mercier et al., 2007). Genetically engineered flat sheeted transdermal capsule implants also termed as "macroencapsulation device" build on polypropylene support had been devised specifically for subcutaneous implantation. It consists of a capsule which can be implanted under skin tissues and expected to release anti-amyloid antibodies over time into systemic circulation, crossing and reaching to the brain. This superficially placed implant shows extended delivery of antibody to meet the intended purpose, that is, can reach to plasma level beyond 50 μg/ml. At the same time, it is able to permeate through brain easily and can bind to existing amyloid plaques. The efficacy testing on TauPS2APP mice presented significant amyloid-β load reduction and consistent release of antibodies prevented the aggregation of amyloids in brain relative to passive immunization by bolus injection. Decreased tau protein phosphorylation was also observed which set a landmark to treat AD (Lathuiliere et al., 2016).

3.3.2 INVASIVE TECHNIQUES FOR CNS TARGETING

Brain targeting is accompanied with a major challenge of overcoming the aforesaid barriers which is a hindrance to most of the drug molecules and peptides. The conventional systemic drug delivery too does not allow a significant increase in the local availability of desired candidate in particular brain regions since the drug gets widely distributed across the tissues. This might also lead to unwanted toxicity. In this scenario, the invasive approach comes to rescue for neurodegenerative disorders. Various techniques form a part of invasive delivery of biopharmaceuticals to the brain is as discussed below (Lu et al., 2014).

3.3.2.1 INTRACEREBRAL IMPLANTS

Intracerebral implants itself is a huge area in which various implants are placed via surgical procedures into a predefined brain region (Fig. 3.3). A novel patented technique including delivery of small-interfering RNAs (siRNA) consists of a catheter implanted surgically with its discharge portion placed adjacent to the infusion site. This approach can be used for treating AD, PD as well as HD. The siRNA would be delivered at the nucleus basalis of Meynart in the brain for silencing the mRNA associated with BACE1 enzyme synthesis in AD while in PD, it interferes with the α-synuclein protein formation when delivered into substantia nigra. For treating the HD the catheter needs to be implanted in such a way that its discharge portion lies adjacent to the striatum.

Polymer-based delivery systems are formulated for treating major brain-related disorders (Govender et al., 2017). Since PD is characterized by low DA levels, combating the DA deficiency using various delivery mechanisms is the approach involved in these polymeric systems. Ethylene vinyl acetate copolymer (EVAc) biocompatible matrix embedded with DA was formulated which was further modified by additional coat of impermeable EVAc with an aperture to achieve constant release pattern of DA. The formulation was further optimized and tested in vivo in rats which resulted in stable DA levels in the brain extracellular fluid for 20 days and the system remained stable for up to 65 days (During et al., 1989; Freese et al., 1989).

Apart from DA treatment, regeneration of dopaminergic neurons with cell-grafting technique in striatum can also result in improved DA levels. Glial-cell-line-derived neurotrophic factor (GDNF) represents factors that protect and regenerate these DA releasing neurons. However due to wide distribution of the GDNF receptors, specificity remains a challenge along with neurological side effects. A polymer system comprising of poly(lactic-*co*-glycolic acid), that is, PLGA microspheres along with GDNF was formulated which was injected into rat striatum in the form of a suspension. The suspending medium for the same constituting distilled water, Tween80, and mannitol. The microspheres were formed by solvent evaporation of O/W/W type emulsion and they showed GDNF entrapment of 84%. The release of GDNF continued up to two months and it led to stimulation of dopaminergic fibers with synaptogenesis. Improvement in behavior and cognitive function were observed (Aoi et al., 2000; Jollivet et al., 2004). The microspheres optimization and premature release of GDNF due to suspension in the mentioned medium remain as shortcomings for this delivery.

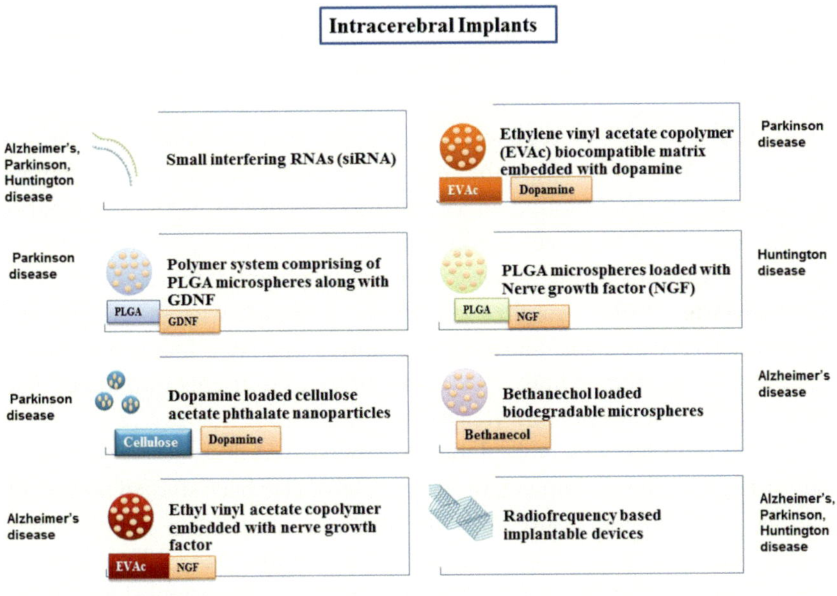

FIGURE 3.3 Intracerebral implants of biopharmaceuticals for neurodegerative disorders.

A similar implantable polymeric structure of PLGA microspheres loaded with NGF was formulated for treatment the HD (Menei et al., 2000). It was implanted a week prior to infusion of quinolinic acid resulting in neuroprotection locally against the excitotoxic lesions. In-vitro tests showed release of approximately 50% of entrapped NGF in sustained manner up to five weeks and in-vivo implant showed degradation post three months. It resulted in approximately 40% reduction in quinolinic acid induced lesions.

Further, a combination of DA loaded cellulose acetate phthalate nanoparticles and polymeric scaffold was studied by Pillay et al. (2009). The nanoparticles were placed into alginate scaffold forming a matrix type polymeric system. After in-vitro testing, the in-vivo studies were carried out in Sprague–Dawley rats. The system was implanted into the frontal lobe parenchyma of rats. It showed prolonged release with controlled levels of DA in CSF. Thus, it can be used for treatment of PD.

To achieve site-specific localization of neurotransmitter, bethanechol loaded biodegradable microspheres comprising copolymer poly-bis-*p*(carboxyphenoxy)propane anhydride and sebacic acid in a 50:50 ratio were formulated. On implantation of these into the denervated hippocampal region, which is the main target area for AD, the performance of rats was improved (Howard et al., 1989). In addition, the implant was well tolerated by rats.

Along the lines of regenerating neurons to compensate for the lost neurons in AD, a biocompatible matrix of ethyl vinyl acetate copolymer embedded with NGF delivered a controlled release of NGF over a period of one month. In PC-12 cells, which are NGF sensitized, it showed formation and growth of neurons thus exhibiting its potential in treatment of AD (Powell et al., 1990). Another innovative approach was an encapsulated cell biodelivery implant. The mechanism involved NGF induced growth of genetically engineered human cells via influx of nutrients into the implant which led to release of NGF (Kapp et al., 2017).

Along with the widely explored polymer-based drug-delivery systems, radiofrequency-based implantable devices have also been studied for treating various CNS-related disorders. One of these being EEG radiotelemetry yields high-quality electrocardiograms (Papazoglou et al., 2016). This can not only be used for the treatment purposes but also for establishing mechanisms involved in the neurodegenerative disease of either superficially or deep into the intracerebral region.

3.3.2.2 INTRATHECAL DELIVERY

Intrathecal delivery involves injecting the drug directly into intrathecal space of the spinal cord (Lu et al., 2014). The drug enters the cerebrospinal fluid present in this space and gets distributed to the brain, thereby bypassing the BBB. The increased proximity of CSF toward the brain also permits reduction in the dose of the drug candidate or biological peptides, hence minimizing the systemic toxicity. Many drugs are known to be more potent and safe via intrathecal route. However, in all cases this does not hold true, instead, the intrathecal route can also be a limitation for penetration of biological macromolecules. Lastly, rapid elimination of drug from the CSF serves as another limitation to this route of delivery. Intrathecal delivery is preferred mainly for anticancer drugs used in the treatment of gliomas and for drugs used in treating brain inflammatory diseases. Various formulation strategies such as microparticles and microparticle-mediated delivery, conjugation of polymers with biomolecules, complexation, and polymer encapsulation have been devised for effective delivery via this route (Soderquist and Mahoney, 2010). This route proved to be helpful in overcoming the systemic side effects that resulted from systemic administration of ciliary neurotrophic factor (CNTF) (Aebischer et al., 1996). Genetically engineered baby hamster kidney cells modified to release human CNTF were encapsulated in a polymer capsule and implanted into the intrathecal space of six ALS patients. These implants showed release of 0.5-μg CNTF/day in vitro. This ex-vivo gene therapy approach did not show any limiting side effects unlike the systemic delivery.

3.3.2.3 BBB DISRUPTION STRATEGIES

To escape the BBB, various strategies for bypassing or disrupting the BBB are depicted in Figure 3.4 and described below.

3.3.2.3.1 Convection-Enhanced Delivery

In this delivery route, the drug formulation is infused directly into brain parenchyma with the aid of a catheter. This system comprises of a pump to create the desired pressure difference for the infusion, a catheter that is inserted at a predefined site in the brain, and the infusate formulation to be delivered at that site. Major advantage of such a delivery is localization of the formulation and reduced volume required to produce the desired pharmacological effect (Lam et al., 2011).

In HD, there is a mutation of the HTT (Huntington) gene. The mHTT (mutant form) involves rapid expansion of the cytosine-adenine-guanine (CAG) repeating domains in the gene. Silencing mRNA of mHTT gene via artificial miRNA will inhibit its translation and thus lead to decreased

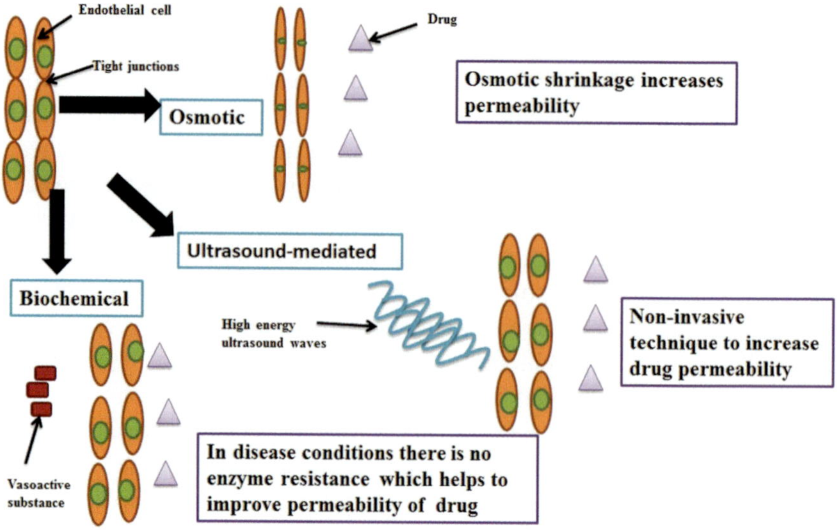

FIGURE 3.4 Disruption strategies for BBB.

mHTT mRNA and/or mHTT protein levels. This has been already studied in mice, however, the reproducibility of these results is essential in humans and hence the effectiveness of this technique in primates closely related to humans was evaluated. The brain size of these animals is relatively larger than that of mice and rats. Here the major challenge is the delivery of artificial miRNA so as to ensure appropriate localization and spreading of the infusate required for treating the disease. The convection-enhanced delivery (CED) technique to deliver artificial miRNA with the help of adeno-associated virus (AAV) vector was evaluated in sheep and the results revealed 50–80% reduction in mHTT RNA and mutant Huntington protein, one and six months postinjection (Pfister et al., 2018). The addition of gadolinium to the delivered infusate helped in tracing the spread of infusate after infusion. The procedure involves lumbar puncture to collect the cerebrospinal fluid. For the infusion, curvilinear incision is made at the desired site followed by drilling of burr hole, which helps in securing the CED cannula in place. Infusion is given after 5 min of cannula insertion with an infusion rate of 3.33 μl/min until 300 μl of total volume is injected. A similar CED route of delivery was used to deliver siRNA in female rhesus monkeys (Stiles et al., 2012). The infusate occupied almost the entire striatum when injected for seven consecutive days. Infusion rates of 0.1, 0.3, and 0.5 μl/min and siRNA concentrations of 4, 8, 12, and 16 mg/ml were investigated. It was found that the threshold suppression concentration of 0.65 mg_{siRNA}/g_{tissue} or greater was required to obtain 46.1% normalization value with 90% confidence level. Minimum 45% normalization was necessary to bring about HTT mRNA suppression corresponding to improved condition. PD treatment with GDNF showed lower volume of distribution since GDNF binds with heparin receptors present in large quantities in the extracellular matrix. To overcome this, CED along with co-infusion of heparin, while administering GDNF as well as NTN (neurturin) was studied. The results showed substantial rise in the volume of distribution of GDNF and when heparin was co-infused with NTN, it improved the DA utility. This led to enhancement in delivery of these two molecules (Hamilton et al., 2001).

3.3.2.3.2 Osmotic BBB Disruption Strategy

In this strategy, the BBB is disrupted by shrinkage of ECs forming cerebral vasculature. This is brought about by administering hyperosmotic solutions of agents, such as mannitol, saline, urea, arabinose, and so on (Lu et al., 2014). In the presence of a hyperosmotic solution, the ECs begin to shrink due to the flow of solvent from inside of the cells to outside thus loosening the gap between these cells ultimately resulting in improved permeability of drugs. However, this technique needs to be carefully executed since there are high chances of neurotoxicity. It was used to deliver hsiRNA (hydrophobically modified siRNA) transvascularly to the rat brain for silencing Htt mRNA (Godinho et al., 2018). Hyperosmotic solution containing 25% mannitol was administered via intracarotid route. The hsiRNA was conjugated with phosphocholine-docosahexanoic acid (PC-DHA) and this technique led to broader distribution of PC-DHA. One week postadministration, the silencing was found to be 55% in striatum, 51% in hippocampus, 52% in somatosensory cortex, 37% in motor cortex, and 33% in thalamus. Apart from mild gliosis, it showed no neurotoxic effects.

3.3.2.3.3 Biochemical BBB Disruption Strategy

In this technique, few molecules which are vasoactive and enhance the permeability of BBB naturally are used. The tendency of strong resistance to these agents by the cells lining the brain capillaries via enzyme secretion, normally acts as barrier for permeability. However in diseased condition, the sites which are affected by the existing disease have less or no resistance for these vasoactive agents. Thus, the permeability is easily improved. An added advantage of this technique is improved permeability at the affected site. This strategy is more reliable and less toxic unlike the osmotic BBB disruption (Lu et al., 2014).

3.3.2.3.4 Ultrasound-Mediated BBB Disruption Strategy

The ultrasonic waves can also be used to enhance BBB permeability. Initially, this technique involved the use of ultrasound at predefined brain area of interest known as focused ultrasound (FUS). It was a noninvasive technique which required high intensity waves to be transmitted that led to heating of the skull and nonuniform distribution of waves due to irregular skull shape, size, and beam aberration. To combat this, FUS thermal ablation technique was invented which led to cooling of the scalp and it comprises a broad aperture transducer that led to wave transmission over a larger area. Later, it was found that prior to the application of ultrasonic waves, an intravenous administration of microbubbles contrast agent would aid in focusing the applied waves at BBB junction. The microbubbles comprise shells made up of lipids or albumin filled with gas such as perfluorocarbon. On ultrasonication, these bubbles oscillate and can increase in size due to diffusion. Further, there can be shear forces arising due to microstreaming of fluid surrounding the microbubbles. These effects result in loosening of tight junctions of the ECs lining brain vasculature. In addition, the transcellular pathways within the cells are activated thereby improving BBB permeability (Aryal et al., 2014).

3.3.3 PASSIVE TARGETING WITH COLLOIDAL DRUG CARRIERS FOR CNS

Colloidal drug carriers include various nanodelivery systems such as liposomes, nanoparticles, nanogels, dendrimers, micelles, nanoemulsions, and quantum dots with size ranging from 1 to 1000 nm. Among these, liposomes and nanoparticles have been mostly explored for brain drug delivery to improve the bioavailability of drugs by enhancing their translocation across

BBB or by increasing the specificity using a targeting ligand. Further, surface functionalization of these carriers with targeting ligands offers an additional advantage of targeting the biopharmaceuticals (Garcia-Garcia et al., 2005).

3.3.3.1 NANOPARTICLES

In AD, there is a rapid loss of cholinergic neurons which is a determinant cause of dementia. NGF regulates the growth, propagation, and overall production of few neurons. Thus, NGF may have potential therapeutic effect in preventing deterioration of dopaminergic neurons and slowing the progression of PD. NGF has a poor BBB penetration hence the carrier system will determine its use in CNS therapy. NGF loaded, polysorbate-80 coated PBCA (polybutyl cyanoacrylate) nanoparticles were formulated using anionic polymerization followed by freeze drying. These nanoparticles were evaluated on 1-methyl-4-phenyl-1,2,3,6-tetrahydropyridine (MPTP)-induced mice and it was found that locomotor activity was increased by 1.9 folds. Further, other symptoms of Parkinsonism such as rigidity, tremors, oligokinesia, and extrapyramidal effects were improved. The passive avoidance reflex test showed reduction in scopolamine-induced amnesia and improvement in recognition and memory. The efficiency of transport across BBB was measured by determination of NGF concentrations in murine brain which demonstrated that the carrier system used is effective when administered intravenously. Thus, it is postulated that this approach may be helpful in improving NGF therapy for age related neurodegenerative disorders (Kurakhmaeva et al., 2009). Urocortin (UN) peptide, which is a corticotropin releasing factor was incorporated in odorranalectin (OL) conjugated PEG–PLGA nanoparticles (OL-NP/UN). The method used for preparation was double emulsion and solvent evaporation. Intranasal administration of these nanoparticles enhanced delivery of drug to brain by crossing BBB. OL-NP/UN exhibited 9.1-fold increase in DA level, as well as decreased the rotational behavior in rats, thus assisting in the treatment of PD (Wen et al., 2011). GDNF gene was incorporated in lactoferrin conjugated nanoparticles (GDNF-Lf NPs), which target the lactoferrin receptors in brain. Lactoferrin has a good BBB penetration and is effective in brain targeting. A single injection of GDNF-Lf NPs showed 3.67 folds increase in the GDNF expression and five injections showed 6.25 folds increase in GDNF expression. The use of multiple injections prolonged the effect of gene expression. It also exhibited an increase in DA level with increase in the monoamine neurotransmitter level and enhanced locomotor activity in PD rats. The study demonstrates to maintain transfection efficiency of nonviral gene vectors and has implications in noninvasive drug delivery (Huang et al., 2010). Metal ion homeostasis is involved in the pathogenesis of AD, which leads to the precipitation and formation of toxic Aβ oligomers. In this case, metal chelating compounds is a promising approach. The metal chelators are not used often due to their severe adverse effects and low penetration in the BBB leading to substantial decrease in bioavailability. Hence, metal chelators were prepared by conjugation with nanocarriers. Polystyrene nanoparticles were conjugated with deferiprone chelators and in-vitro study on human cortical neurons showed prevention of Aβ aggregation. In another in-vitro study, copper chelator, D-penicillamine coupled with hexadecanol nanoparticles also prevented Aβ aggregation. Zinc and copper chelator, clioquinol, incorporated into the PBCA nanoparticles was evaluated in AD transgenic mice. It proficiently crossed BBB and also reduced the signs of AD (Hadavi and Poot, 2016).

3.3.3.2 LIPOSOMES

Liposomal vaccine was prepared by incorporating Aβ peptide on the surface of liposomes by coupling two fatty acid residues or phospholipids with PEG at each end of peptide. The Aβ peptide formed β-sheet conformation due to interaction of peptide with liposomal surface. When PEG spacers were used, a random coil-like conformation was seen in the liquid phase of liposome. Intraperitoneal administration in mice showed decrease in cognitive function and a significant reduction in soluble as well insoluble Aβ peptide in the brain of mice. This data showed that AD can be treated with vaccines against β-sheet conformations of Aβ peptide (Lin et al., 2016).

Liposomes containing plasmid DNA-microbubble complex (LpDNA-MBs) were prepared. Lp-DNA consisted of green fluorescence protein (GFP) and GDNF. FUS triggered LpDNA-MBs exhibited successful opening of the BBB and thereby increased expression of GFP and GDNF proteins. Increase in the level of DA and its metabolite DOPAC, HVA was seen. Thus, the degradation of DA receptors was prevented resulting in the restoration of DA in the PD-mice model. It also helped in reversing motor behavior and improving neuronal functions. Thus, it is a potential approach for various neurodegenerative disorders (Gonzalez-Barrios et al., 2006). Neurotensinpolyplex was used as a carrier system for human glial cell-line-derived neurotrophic factor (hGDNF). Neurotensin and poly-L-Lysine were cross-linked and appropriate gel filtration was done to purify the obtained derivatives. Neurotensinpolyplex was made by electrostatically binding the mutant with pDNA. This hGDNF was transfected for one week in substantia nigra of rats and after neurotoxin injection, reversal of motor impairment with 90% and 50% recovery of DA level in striatum and substantia nigra, respectively was observed. It also assisted in biochemical and functional recovery. This study showed that neurotensin-polyplex holds the potential of targeting the reporter gene of dopaminergic neurons in the nigral region (Garbayo et al., 2012).

3.3.3.3 STEM CELL THERAPY

Cell transplantation is an efficient therapy for the treatment of neurodegenerative disorders. However, cell therapy is not preferred due to delivery issues and to overcome these constraints PLGA microparticles were formulated. PLGA conjugated pharmacologically active microparticles (PAMs) that mimic the biological surface and provide controlled release of cells and growth factor were prepared. This biomimetic property helps in growth and survival of cells. First, the NGF-releasing PAMs (NGF-PAMs) were transported to neuronal cell line PC12 cells and then GDNF-releasing PAMs were targeted to dopaminergic cells. The PLGA–PAMs can be implanted into any region of brain by stereotaxic surgery and this would help in tissue repair of the brain in case of neurodegenerative disorders (Garbayo et al., 2012).

3.3.3.4 MRNA-LOADED NANOMICELLES TARGETING BRAIN

The approach involving breaking of equilibrium between soluble and insoluble Aβ is an effective treatment for AD. Neprilysin plays a vital role in the clearance of Aβ and its conjugation with viral vectors has demonstrated decrease in Aβ deposition and prevention of pathogenic changes in the brain.

mRNA-loaded polyplex nanomicelles were prepared which showed higher expression of gene in the brain upon intracerebroventricular injection.

The major concern with mRNA is its stability and strong immunogenicity. However, these complications were overcome with nanomicelles effectively. Neprilysin-loaded mRNA nanomicelles were infused in AD-induced mice and they showed decrease in β-amyloid protein, thus proving that mRNA-loaded nanomicelles can efficiently deliver proteins and peptides to the brain (Lin et al., 2016).

3.3.4 ACTIVE TARGETING FOR CNS

Active targeting is a noninvasive technique for traversing BBB. In this technique, a therapeutic moiety is conjugated with a ligand capable of recognizing and binding to a specific target site. Thus, this conjugation imparts affinity to the therapeutic moiety towards its target. These targets include various transport arrangements shown within the cerebral ECs. The ligand here is not a drug but a facilitator to enhance the drug delivery. Over 20 transporters have been recognized, all highly expressed on the cerebral capillaries of BBB. Various approaches have been identified for active targeting to BBB which include receptor/vector-mediated delivery, cell-penetrating peptides (CPP), and viral vectors (Fig. 3.5) (Beduneau et al., 2007).

3.3.4.1 RECEPTOR/VECTOR-MEDIATED DELIVERY

Receptor-mediated transport mechanisms involve vesicular transport system of the brain endothelium. Transcytosis is nothing but the influx of nutrients followed by a transcellular, receptor-mediated transport (Jones and Shusta, 2007). Precisely, a circulating ligand binds with a particular receptor of EC surface. After ligand binding, the process of endocytosis begins, which leads to the formation of intracellular transport vesicles. These vesicles containing receptor-ligand complexes are sent to the basolateral side of EC, where they get released. In this way, molecules can cross BBB (Wang et al., 2009). Such transport can either occur via receptor-mediated transcytosis (RMT) or AMT (Xiao and Gan, 2013). RMT involves initial binding of ligand to an entity on or inside plasma membrane of the ECs, whereas AMT is based upon nonspecific charge based interactions that can be initiated by binding of polycationic molecules to the plasma membrane due to negative charges (Lu, 2012;

Herve et al., 2008). But due to lack of specificity, AMT is not much used. Over the years, RMT has developed as a class of novel system for delivery through BBB; various specific receptors have been identified on the brain capillaries which include insulin (Ulbrich et al., 2011; Lajoie and Shusta, 2015), IGF-1 and 2 (Bake et al., 2016), angiotensin 2 (Rose and Audus., 1998), transferrin (Johnsen et al., 2017), and so on. For utilizing RMT for delivery of drugs to brain, the therapeutic moiety must be attached to a molecule capable of targeting the RMT system.

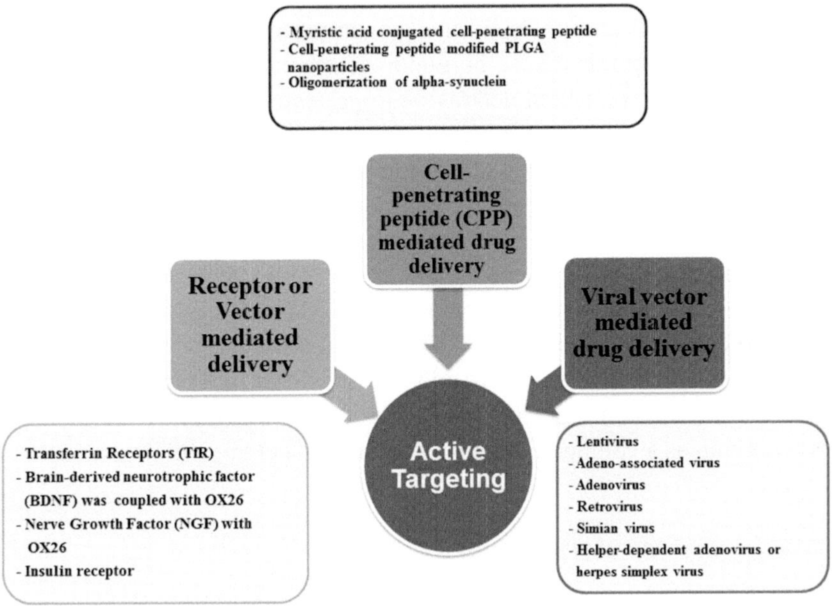

FIGURE 3.5 Active targeting of biopharmaceuticals for CNS.

Transferrin receptors (TfR) have been exploited frequently for designing drug delivery to brain due to their high expression on the luminal side of the brain capillary endothelium. Activation of these receptors leads to RMT through BBB (Ulbrich et al., 2009). The mouse monoclonal antibody (MAb) against the rat TfR, OX26, binds to a distinct TfR binding site preventing competition for binding site between the vector and naturally occurring Tf. It has thus been extensively studied for varied applications in brain drug delivery. In one of the earliest studies, Saito et al. studied the delivery of monobiotinylated ^{125}I-labeled A β 1-40 through BBB by conjugating it with streptavidin (SA)-OX26, which showed a 2- fold greater brain uptake. Further, binding of ^{125}I, bio-A β 1-40/SA-OX26 conjugate to the amyloid of AD brain was evaluated by film and emulsion autoradiography done on frozen sections of AD brain (Saito et al., 1995). In another study, brain-derived neurotrophic factor (BDNF) that is a neuroprotective agent, was coupled with OX26 (previously functionalized with biotin and SA) yielding an OX26-BDNF conjugate which showed 243% increase in motor performance relative to BDNF alone when injected in rats through permanent middle cerebral artery occlusion.

Kordower et al. (1994) conjugated NGF with OX26 and studied it in rat model of HD and observed the prevention of deterioration of striatal cholinergic acetyltransferase-immunoreactive neurons. Human bFGF, a neurotrophic factor which acts as neuroprotective in conditions of brain ischemia, which when conjugated with OX26-(SA/B) and delivered intravenously showed 80% reduction in infarct volume (Song et al., 2002). Similar to TfR, insulin receptor is also observed on luminal membrane of the brain capillary EC as well as in the plasma membrane of other brain cells which can undergo RMT across BBB. The 83-14 mouse MAb has been used against human insulin receptor as an RMT delivery vector. In one of the studies, a radiolabeled $A\beta^{1-40}$ 83–14 fusion protein was synthesized to trace the amyloid burden in vivo and when injected into rhesus monkey via IV route, the concentration of radiolabeled $A\beta^{1-40}$ was found to be enhanced when given as a conjugate with 83-14 Mab (Wu et al., 1997).

3.3.4.2 CPP-MEDIATED DRUG DELIVERY

Most of the delivery systems used to transport therapeutic molecules into CNS exhibit various limitations when applied in clinical situations whereas CPP have been projected as a promising delivery system showing great ability in transporting molecules into CNS with low cellular toxicity and high efficiency. CPPs, because of their small size (up to 30 amino acids in length) demonstrate high potential to cross BBB which enables improved delivery of therapeutics to treat CNS diseases. Various CPPs have been employed for this purpose including TAT, TP, rabies virus glycoprotein (RVG), angiopep, penetratin, prion peptide, and SynB (Cai et al., 2011; Fonseca et al., 2009; Gabathuler, 2010; Gallo, 2003; Kumar et al., 2007; Mae and Langel 2006).

There are two main proposed mechanisms for the action of CPP, which are direct penetration and endocytosis-mediated entry. The earlier one is an energy-independent translocation across cellular membranes involving a direct electrostatic interaction with negatively charged phospholipids. It has been proposed that it involves a strong interaction between CPPs and phosphate groups on both sides of the lipid bilayer (Herce and Garcia, 2007). Second mechanism that is, endocytosis-mediated entry is an energy-dependent process wherein the plasma membrane folds inward and the cell absorbs surrounding substances by imbibing them within their cellular membrane. Studies have also shown that this process was induced by interaction of polyarginines with heparan sulfates for cellular entry of penetratin (Frankel and Pabo, 1988; Lundberg and Johansson, 2001). In another study, the penetrating dimer was conjugated with negatively charged phospholipids, thus generating an inverted micelle inside the lipid bilayer. This inverted micelle formation enhances the delivery of peptide (Deshayes et al., 2004; Plenat et al., 2004). Apart from these two basic mechanisms, several other mechanisms have been proposed for the internalization of CPP such as micropinocytosis, clathrin-mediated endocytosis, and caveolae-mediated endocytosis. It was also shown that the cationic conjugates rely on micropinocytosis, whilst amphipathic conjugates utilize clathrin-mediated endocytosis (Jiao et al., 2009). Recently, effective delivery and functional release of bioactive siRNA has been proposed as a promising approach to treat neurodegenerative disorders. Youn et al. (2014), synthesized a myristic acid conjugated CPP, transportan (TP), prepared with transferrin receptor-targeting peptide (myr-TP-Tf). A successful siRNA uptake was achieved as indicated by fluorescence images proposing myr-TP-Tf as a promising vehicle for neurotargeted siRNA delivery. In an interesting study, CPP-modified PLGA nanoparticles were synthesized for the delivery of insulin to brain via intranasal route as a potential therapy against AD. The results indicated that CPP can efficiently deliver insulin into brain via the nasal route with an efficiency of 6% (Yan et al., 2013).

The intracellular oligomerization of alpha-synuclein is an important disease-modifying treatment associated with PD and to exploit this mechanism, stable cell penetrating β-synuclein fragments have been identified. Among these peptides, β-syn 6short and β-syn 37 inhibit fibril formation but not oligomer formation in contrast to β-syn 78, which inhibits oligomer formation but not fibril formation. The β-syn 36 peptide inhibited both the soluble and insoluble fibrillar aggregates of a-syn with no cytotoxicity (Shaltiel-Karyo et al., 2010).

3.3.4.3 VIRAL VECTOR-MEDIATED DRUG DELIVERY

Another way of active targeting is a viral-mediated drug transfer that can overcome the restrictions for transporting the molecules into the brain through BBB. Due to their regular tendency of crossing the cellular membrane to infect postmitotic cells and to transfer genetic material into target cells, viral vectors derived from lentivirus, AAV, adenovirus (Ad), retrovirus, simian virus, helper-dependent Ad, or even herpes simplex virus were used. They have been explored as gene delivery vehicles either by directly injecting the viral vector into cerebral lateral ventricles or at multiple sites (Janson et al., 2001; Davidson and Breakefield, 2003; Danos and Mulligan, 1988; Emi et al., 1991). Further, mannitol (McCarty et al., 2009) and heparin (Woodard et al., 2016) have been used for osmotic disruption of the BBB and enhance the delivery of viral vectors.

Retroviruses are enveloped viruses carrying diploid RNA genome of about 8–10 kb. The virus binds to a specific cell-surface receptor leading to the internalization of the particle. This further causes uncoating of the virus in cytoplasm wherein the reverse transcriptase converts genomic RNAs into double-stranded DNA molecules that are carried to nucleus and subsequently incorporated into the genome (Roe et al., 1993; Lewis and Emerman, 1994). Lentiviruses belong to retrovirus family obtained from human immunodeficiency virus 1 or nonhuman viruses (feline and simian immunodeficiency virus, and equine infection anemia virus) having potential to translocate through nuclear membrane (Mitrophanous et al., 1999; Poeschla et al., 1998).

Other accessory and regulatory genes associated with lentiviruses could probably cause harmful effects if transfected as they are biologically active. Hence, minimal vectors that lack lentiviral Tat gene such as Rev are synthesized and only lentiviral proteins being expressed are gag/pol. These have been a major breakthrough in lentivirus-mediated gene delivery (Mazarakis et al., 2001). GDNF have been used as a neuroprotective agent for PD which when delivered using lentivirus efficiently prevented loss of dopaminergic nigrostriatal neurons in rodent and monkey models of PD (Bensadoun et al., 2000; Kordower et al., 2000). To explore a novel approach to treat HDF; CNTF-expressing lentiviral vectors were investigated using quinolinic acid model of HD and it was demonstrated that the lesion volume was significantly reduced when compared with control group (de Almeida et al., 2001).

Lentiviral vectors carrying siRNA against BACE1 were synthesized and investigated in APP transgenic mice wherein it was observed that reduction in level of BACE1 resulted in enhanced spatial memory and learning in Morris water maze. In addition, reduction in amyloid deposition and decrease in neuronal loss in hippocampus suggested that lentiviral vector is an encouraging tool for the treatment of AD (Singer et al., 2005). AAV comprises of around 4.7 kb genome and 20–24 nm in diameter is a small non enveloped virus containing single-stranded DNA molecule which postadministration gets transformed to a double-stranded template (Rabinowitz and Samulski, 1998; Snyder, 1999). In the various AAV serotypes available, AAV2 is the most commonly used to target brain (Bartlett et al., 1998). However, AAV9 was used in PD to deliver human erythropoietin into the rat striatum (Xue et al., 2010). In addition to these, AAV serotypes 2, 4, and 5 have been exploited to develop vector capsids (Van Vliet et al., 2008). Apart from this, it is interesting to note that AAV5 and AAV1 transduce both neurons and glial cells, whereas AAV4 transduces ependymal cells predominantly (Davidson et al., 2000; Wang et al., 2003).

Recombinant AAV have been investigated as a gene delivery tool to deliver rat GDNF in 6-hydroxydopamine-lesioned rats wherein it was demonstrated that nigral dopaminergic neurons were unaltered as compared to control. On similar bases in different studies AAV2 and AAV5 have been used to deliver GDNF wherein reduction in ubiquitin-proteasome system-induced degeneration of the dopaminergic neurons was observed accompanied by neurogenesis in the subventricular zone and dentate gyrus of the brain (Du et al., 2013; Tereshchenko et al., 2014). In another interesting study, recombinant AAV-GDNF was delivered along with grafted DA precursor cells into the putamen and enhancement in DA level was observed (Redmond et al., 2013). Cederfjall et al. (2012) synthesized a single AAV vector coexpressing TH and GTPCHI for DA secretion in Sprague–Dawley rats wherein data revealed supranormal and subnormal DA concentrations in postsynaptic region. In further studies, to explore a novel approach for controlled DA reconstitution in the brain of PD patients, Cederfjall et al. (2013), used TH expressing AAV to control the expression of dihydrofolate reductase-guanosine triphosphate cyclohydrolase I (DD-GTPCHI) across the BBB (Cederfjall et al., 2015).

There are various therapeutic agents under clinical investigation for neurodegenerative disease and are summarized in Table 3.1.

TABLE 3.1 Clinical Trials Conducted for Neurodegenerative Disease.

S. No.	Therapeutic agents	Sponsor/Company	Phase	Clinical trials reg. ID	Reference
			Alzheimer's disease		
1.	Solanezumab	Eli Lily	Phase III—terminated	NCT02760602	https://clinicaltrials.gov/
2.	Ganteneumab	Roche	Phase III—ongoing	NCT03444870	
3.	Crenezumab	Genetech/Roche	Phase III—ongoing	NCT02353598	
4.	Bapineuzumab	Elan/Pfizer and Johnson and Johnson	Phase II—terminated	NCT00667810	
5.	Aducanumab	Biogen	Phase III—ongoing	NCT02477800	
6.	Ponezumab	Pfizer Inc.	Phase IIa—terminated	NCT01821118	
7.	BAN2401	Elsa/BioArctic Neuroscience	Phase IIb—ongoing	NCT01767311	
8.	SAR228810	Sanofi	Phase I—completed	NCT01485302	
9.	GSK933776A	GSK	Phase I—completed	NCT00459550	
10.	RN6G (Pf-04382923) IgG2	Pfizer (Rinat Neuroscience)	Phase II—completed	NCT01577381	
11.	LY3002813 IgG1 (with BACE inhibitor LY3202626)	Eli Lilly	Phase II—ongoing	NCT03367403	
12.	MEDI1814	AstraZeneca	Phase I—completed	NCT02036645	
			Parkinson's disease		
13.	AFFITOPE PD01A	Affiris AG	I	NCT01885494	
14.	BIIB054	Biogen	I	NCT02459886	
15.	PRX002	Prothena Biosciences	I	NCT02157714	
16.	AFFITOPE PD03A	Affiris AG	I	NCT02267434	
17.	Intranasal insulin	University of Southern California	II	NCT01767909	
18.	Intranasal insulin	Health Partners Institute	II	NCT02503501	
19.	Exenatide	University College, London	II	NCT01971242	

TABLE 3.1 *(Continued)*

S. No.	Therapeutic agents	Sponsor/Company	Phase	Clinical trials reg. ID	Reference
20.	AADC (L-amino acid decarboxylase)	AADC (L-amino acid decarboxylase)	I	NCT00229736	
21	GAD (glutamic acid decarboxylase)	Neurologix, Inc.	II	NCT0064390	
22.	GDNF (glial cell-line derived neurotrophic factor)	North Bristol NHS Trust	II	NCT03652363	
23.	NRTN (neurturin)	Ceregene	I	NCT00252850	
		Multiple sclerosis			
24.	Siponimod	Novartis	III	NCT01665144	
25.	Ozanimod	Celgene	III	NCT02576717	
26.	High-dose biotin	MedDay Pharmaceuticals	III	NCT02220244	
27.	Secukinumab	Novartis	II	NCT01874340	
		Huntington's disease			
28.	IONIS-HTT_{RX} OLE	Ionis Pharmaceuticals Inc.	II	NCT03342053	
29.	PRECISION-HD1	Wave Life Sciences Ltd.	I	NCT03225833	
30.	PRECISION-HD2	Wave Life Sciences Ltd.	I	NCT03225846	
31.	SIGNAL	Vaccinex Inc., Huntington Study Group	II	NCT02481674	
32.	LEGATO-HD	Teva Branded Pharmaceutical Products, R&D Inc.	II	NCT02215616	
33.	ADORE-HD	AzidusBrasil	II	NCT03252535	
34.	BMACHC	Chaitanya Hospital, Pune	I/II	NCT01834053	

3.4 CONCLUSION

With the development of the modern society and the emerging concerns regarding the deterioration of environment, there are increased incidences of neurodegenerative disorders like AD and PD that can be devastating for human well-being.

For the treatment of these CNS disorders, the challenging part is overcoming physiological barriers and maintaining therapeutic concentrations at the target sites for a prolonged period. BBB and BCSFB restrict the entry of drugs due to the presence of continuous tight junctions and enzymes. To have effective CNS delivery of biopharmaceuticals, site-specific targeted drug delivery is of prime importance. The key features affecting the effectiveness of drug delivery to CNS include physicochemical properties of drug, drug transport mechanisms, its pharmacokinetics, and also its ability to cross the physiological barriers.

The noninvasive delivery of biopharmaceuticals to CNS has been reported via transdermal and intranasal route and is well accepted due to its ease of administration. The invasive techniques that have been adopted for CNS targeting of biopharmaceuticals include intrathecal and intracerebral delivery. intrathecal delivery, targeting is achieved with the use of polymeric and microparticulate systems. The effective targeting in intracerebral route is achieved with the use of nanoparticulate, polymeric, microspheres, and radiofrequency-based approaches. The BBB disruption strategies have also been used to ease the delivery of biopharmaceuticals in the brain. These strategies are based upon osmotic approach, ultrasound-mediated, or biochemical approach. The active and passive targeting approaches have also been used for improving the drug access at the target site. Various biotechnological products have been employed in CNS targeting due to their ability in achieving site-specific targeting and scope for easy structural modifications.

To conclude, CNS delivery of biopharmaceuticals is an expanding and challenging field. Designing formulation for CNS targeting requires a sound understanding of physicochemical properties of drug, pharmacokinetic parameters, and biological parameters. The clarity on molecular mechanisms of BBB and BCSFB shall provide knowledge in designing more efficacious formulations which are site-specific.

KEYWORDS

- **biopharmaceuticals**
- **neurodegenerative disease**
- **drug delivery**
- **active targeting**

REFERENCES

Aebischer, P.; Schluep, M.; Deglon, N.; Joseph, J. M.; Hirt, L.; Heyd, B.; Goddard, M.; Hammang, J. P.; Zurn, A. D.; Kato, A. C.; Regli, F.; Baetge, E. E. Intrathecal Delivery of CNTF Using Encapsulated Genetically Modified Xenogeneic Cells in Amyotrophic Lateral Sclerosis Patients. *Nat. Med.* **1996,** *2* (6), 696–699.

Alavijeh, M. S.; Chishty, M.; Qaiser, M. Z.; Palmer, A. M. Drug Metabolism and Pharmacokinetics, the Blood-Brain Barrier, and Central Nervous System Drug Discovery. *NeuroRx* **2005,** *2* (4), 554–571.

Albarran, B.; Hoffman, A. S.; Stayton, P. S. Efficient Intracellular Delivery of a Pro-apoptotic Peptide with a pH-Responsive Carrier. *React. Funct. Polym.* **2011,** *71*, 261–265.

Aloe, L.; Rocco, M. L.; Bianchi, P.; Manni, L. Nerve Growth Factor: From the Early Discoveries to the Potential Clinical Use. *J. Transl. Med.* **2012,** *10*, 239.

Alzheimer's, A. Alzheimer's Disease Facts and Figures. *Alzheimer's Dement* **2015,** *11* (3), 332–384.

Antonini, A.; Abbruzzese, G.; Barone, P.; Bonuccelli, U.; Lopiano, L.; Onofrj, M.; Zappia, M.; Quattrone, A. COMT Inhibition with Tolcapone in the Treatment Algorithm of Patients with Parkinson's Disease (PD): Relevance for Motor and Non-motor Features. *Neuropsychiatr. Dis. Treat.* **2008,** *4* (1), 1–9.

Aoi, M.; Date, I.; Tomita, S.; Ohmoto, T. The Effect of Intrastriatal Single Injection of GDNF on the Nigrostriatal Dopaminergic System in Hemi-Parkinsonian Rats: Behavioral and Histological Studies Using Two Different Dosages. *Neurosci. Res.* **2000,** *36* (4), 319–325.

Aryal, M.; Arvanitis, C. D.; Alexander, P. M.; McDannold, N. Ultrasound-Mediated Blood–Brain Barrier Disruption for Targeted Drug Delivery in the Central Nervous System. *Adv. Drug Deliv. Rev.* **2014,** *72*, 94–109.

Avgerinos, K. I.; Kalaitzidis, G.; Malli, A.; Kalaitzoglou, D.; Myserlis, P. G.; Lioutas, V. A. Intranasal Insulin in Alzheimer's Dementia or Mild Cognitive Impairment: A Systematic Review. *J. Neurol.* **2018,** *265* (7), 1497–1510.

Bake, S.; Okoreeh, A. K.; Alaniz, R. C. Sohrabji, F. Insulin-Like Growth Factor (IGF)-I Modulates Endothelial Blood–Brain Barrier Function in Ischemic Middle-Aged Female Rats. *Endocrinology* **2016,** *157* (1), 61–69.

Ballabh, P.; Braun, A.; Nedergaard, M. The Blood–Brain Barrier: An Overview: Structure, Regulation, and Clinical Implications. *Neurobiol. Dis.* **2004,** *16* (1), 1–13.

Bano, D.; Zanetti, F.; Mende, Y.; Nicotera, P. Neurodegenerative Processes in Huntington's Disease. *Cell Death Dis.* **2011,** *2*, e228.

Bartlett, J. S.; Samulski, R. J.; McCown, T. J. Selective and Rapid Uptake of Adeno-Associated Virus Type 2 in Brain. *Hum. Gene Ther.* **1998,** *9* (8), 1181–1186.

Beduneau, A.; Saulnier, P.; Benoit, J. P. Active Targeting of Brain Tumors Using Nanocarriers. *Biomaterials* **2007,** *28* (33), 4947–4967.

Bensadoun, J. C.; Deglon, N.; Tseng, J. L.; Ridet, J. L.; Zurn, A. D.; Aebischer, P. Lentiviral Vectors as a Gene Delivery System in the Mouse Midbrain: Cellular and Behavioral Improvements in a 6-OHDA Model of Parkinson's Disease Using GDNF. *Exp. Neurol.* **2000,** *164* (1), 15–24.

Bingham, A.; Ekins, S. Competitive Collaboration in the Pharmaceutical and Biotechnology Industry. *Drug Discov. Today* **2009,** *14* (23–24), 1079–1081.

Cai, B.; Lin, Y.; Xue, X. H.; Fang, L.; Wang, N.; Wu, Z. Y. TAT-Mediated Delivery of Neuroglobin Protects against Focal Cerebral Ischemia in Mice. *Exp. Neurol.* **2011,** *227* (1), 224–231.

Capsoni, S.; Ugolini, G.; Comparini, A.; Ruberti, F.; Berardi, N.; Cattaneo, A. Alzheimer-Like Neurodegeneration in Aged Antinerve Growth Factor Transgenic Mice. *Proc. Natl. Acad. Sci. U. S. A.* **2000,** *97* (12), 6826–6831.

Carvey, P. M.; Hendey, B.; Monahan, A. J. The Blood–Brain Barrier in Neurodegenerative Disease: A Rhetorical Perspective. *J. Neurochem.* **2009,** *111* (2), 291–314.

Cederfjall, E.; Sahin, G.; Kirik, D.; Bjorklund, T. Design of a Single AAV Vector for Coexpression of TH and GCH1 to Establish Continuous DOPA Synthesis in a Rat Model of Parkinson's Disease. *Mol. Ther. J. Am. Soc. Gene Ther.* **2012,** *20* (7), 1315–1326.

Cederfjall, E.; Nilsson, N.; Sahin, G.; Chu, Y.; Nikitidou, E.; Bjorklund, T.; Kordower, J. H.; Kirik, D. Continuous DOPA Synthesis from a Single AAV: Dosing and Efficacy in Models of Parkinson's Disease. *Sci. Rep.* **2013,** *3*, 2157.

Cederfjall, E.; Broom, L.; Kirik, D. Controlled Striatal DOPA Production from a Gene Delivery System in a Rodent Model of Parkinson's Disease. *Mol. Ther.* **2015,** *23* (5), 896–906.

Chaudhuri, A. Multiple Sclerosis Is Primarily a Neurodegenerative Disease. *J. Neural Transm.* **2013,** *120* (10), 1463–1466.

Cheng, Y. S.; Chen, Z. T.; Liao, T. Y.; Lin, C.; Shen, H. C.; Wang, Y. H.; Chang, C. W.; Liu, R. S.; Chen, R. P.; Tu, P. H. An Intranasally Delivered Peptide Drug Ameliorates Cognitive Decline in Alzheimer Transgenic Mice. *EMBO Mol. Med.* **2017,** *9* (5), 703–715.

Covaceuszach, S.; Capsoni, S.; Ugolini, G.; Spirito, F.; Vignone, D.; Cattaneo, A. Development of a Noninvasive NGF-Based Therapy for Alzheimer's Disease. *Curr. Alzheimer Res.* **2009,** *6* (2), 158–170.

Danos, O.; Mulligan, R. C. Safe and Efficient Generation of Recombinant Retroviruses with Amphotropic and Ecotropic Host Ranges. *Proc. Natl. Acad. Sci. U. S. A.* **1988,** *85* (17), 6460–6464.

Davidson, B. L.; Breakefield, X. O. Viral Vectors for Gene Delivery to the Nervous System. *Nat. Rev. Neurosci.* **2003,** *4* (5), 353–364.

Davidson, B. L.; Stein, C. S.; Heth, J. A.; Martins, I.; Kotin, R. M.; Derksen, T. A.; Zabner, J.; Ghodsi, A.; Chiorini, J. A. Recombinant Adeno-Associated Virus Type 2, 4, and 5

Vectors: Transduction of Variant Cell Types and Regions in the Mammalian Central Nervous System. *Proc. Natl. Acad. Sci. U. S. A.* **2000,** *97* (7), 3428–3432.

de Almeida, L. P.; Zala, D.; Aebischer, P.; Deglon, N. Neuroprotective effect of a CNTF-Expressing Lentiviral Vector in the Quinolinic Acid Rat Model of Huntington's Disease. *Neurobiol. Dis.* **2001,** *8* (3), 433–446.

De Rosa, R.; Garcia, A. A.; Braschi, C.; Capsoni, S.; Maffei, L.; Berardi, N.; Cattaneo, A. Intranasal Administration of Nerve Growth Factor (NGF) Rescues Recognition Memory Deficits in AD11 Anti-NGF Transgenic Mice. *Proc. Natl. Acad. Sci. U. S. A.* **2005,** *102* (10), 3811–3816.

Deshayes, S.; Gerbal-Chaloin, S.; Morris, M. C.; Aldrian-Herrada, G.; Charnet, P.; Divita, G.; Heitz, F. On the Mechanism of Non-endosomial Peptide-Mediated Cellular Delivery of Nucleic Acids. *Biochim. Biophys. Acta* **2004,** *1667* (2), 141–147.

Dhuria, S. V.; Hanson, L. R.; Frey, W. H., 2nd. Intranasal Drug Targeting of Hypocretin-1 (Orexin-A) to the Central Nervous System. *J. Pharm. Sci.* **2009,** *98* (7), 2501–2515.

Dong, X. Current Strategies for Brain Drug Delivery. *Theranostics* **2018,** *8* (6), 1481–1493.

Du, Y.; Zhang, X.; Tao, Q.; Chen, S.; Le, W. Adeno-Associated Virus Type 2 Vector-Mediated Glial Cell Line-Derived Neurotrophic Factor Gene Transfer Induces Neuroprotection and Neuroregeneration in a Ubiquitin-Proteasome System Impairment Animal Model of Parkinson's Disease. *Neurodegenerat. Dis.* **2013,** *11* (3), 113–128.

During, M. J.; Freese, A.; Sabel, B. A.; Saltzman, W. M.; Deutch, A.; Roth, R. H.; Langer, R. Controlled Release of Dopamine from a Polymeric Brain Implant: *In Vivo* Characterization. *Ann. Neurol.* **1989,** *25* (4), 351–356.

Emi, N.; Friedmann, T.; Yee, J. K. Pseudotype Formation of Murine Leukemia Virus with the G Protein of Vesicular Stomatitis Virus. *J. Virol.* **1991,** *65* (3), 1202–1207.

Fernando, G.; Yamila, R.; Cesar, G. J.; Ramon, R. Neuroprotective Effects of neuro-EPO Using an *In Vitro* Model of Stroke. *Behav. Sci.* **2018,** *8* (2), 26.

Fonseca, S. B.; Pereira, M. P.; Kelley, S. O. Recent Advances in the Use of Cell-Penetrating Peptides for Medical and Biological Applications. *Adv. Drug Deliv. Rev.* **2009,** *61* (11), 953–964.

Frankel, A. D.; Pabo, C. O. Cellular Uptake of the Tat Protein from Human Immunodeficiency Virus. *Cell* **1988,** *55* (6), 1189–1193.

Freese, A.; Sabel, B. A.; Saltzman, W. M.; During, M. J.; Langer, R. Controlled Release of Dopamine from a Polymeric Brain Implant: *In Vitro* Characterization. *Exp. Neurol.* **1989,** *103* (3), 234–238.

Gabathuler, R. Development of New Peptide Vectors for the Transport of Therapeutic across the Blood-Brain Barrier. *Ther. Deliv.* **2010,** *1* (4), 571–586.

Gallo, G. Making Proteins into Drugs: Assisted Delivery of Proteins and Peptides into Living Neurons. *Methods Cell Biol.* **2003,** *71*, 325–338.

Garbayo, E.; Ansorena, E.; Blanco-Prieto, M. J. Brain Drug Delivery Systems for Neurodegenerative Disorders. *Curr. Pharm. Biotechnol.* **2012,** *13* (12), 2388–2402.

Garcia-Garcia, E.; Andrieux, K.; Gil, S.; Couvreur, P. Colloidal Carriers and Blood-Brain Barrier (BBB) Translocation: A Way to Deliver Drugs to the Brain? *Int. J. Pharm.* **2005,** *298* (2), 274–292.

Ghose, A. K.; Herbertz, T.; Hudkins, R. L.; Dorsey, B. D.; Mallamo, J. P. Knowledge-Based, Central Nervous System (CNS) Lead Selection and Lead Optimization for CNS Drug Discovery. *ACS Chem. Neurosci.* **2012,** *3* (1), 50–68.

Gitler, A. D.; Dhillon, P.; Shorter, J. Neurodegenerative Disease: Models, Mechanisms, and a New Hope. *Dis. Models Mech.* **2017,** *10* (5), 499–502.

Godinho, B.; Henninger, N.; Bouley, J.; Alterman, J. F.; Haraszti, R. A.; Gilbert, J. W.; Sapp, E.; Coles, A. H.; Biscans, A.; Nikan, M.; Echeverria, D.; DiFiglia, M.; Aronin, N.; Khvorova, A. Transvascular Delivery of Hydrophobically Modified siRNAs: Gene Silencing in the Rat Brain upon Disruption of the Blood-Brain Barrier. *Mol. Ther.* **2018,** *26* (11), 2580–2591.

Gonzalez-Barrios, J. A.; Lindahl, M.; Bannon, M. J.; Anaya-Martinez, V.; Flores, G.; Navarro-Quiroga, I.; Trudeau, L. E.; Aceves, J.; Martinez-Arguelles, D. B.; Garcia-Villegas, R.; Jimenez, I.; Segovia, J.; Martinez-Fong, D. Neurotensin Polyplex as an Efficient Carrier for Delivering the Human GDNF Gene into Nigral Dopamine Neurons of Hemi-Parkinsonian Rats. *Mol. Ther. J. Am. Soc. Gene Ther.* **2006,** *14* (6), 857–865.

Govender, T.; Choonara, Y. E.; Kumar, P.; Bijukumar, D.; du Toit, L. C.; Modi, G.; Naidoo, D.; Pillay, V. Implantable and Transdermal Polymeric Drug Delivery Technologies for the Treatment of Central Nervous System Disorders. *Pharm. Dev. Technol.* **2017,** *22* (4), 476–486.

Goyal, K; Koul V.; Singh, Y.; Anand, A. Targeted Drug Delivery to Central Nervous System (CNS) for the Treatment of Neurodegenerative Disorders: Trends and Advances. *Cent. Nerv. Syst. Agents Med. Chem.* **2014,** *14* (1), 43–59.

Hadavi, D.; Poot, A. A. Biomaterials for the Treatment of Alzheimer's Disease. *Front. Bioeng. Biotechnol.* **2016,** *4*, 49.

Hamilton, J. F.; Morrison, P. F.; Chen, M. Y.; Harvey-White, J.; Pernaute, R. S.; Phillips, H.; Oldfield, E.; Bankiewicz, K. S. Heparin Coinfusion During Convection-Enhanced Delivery (CED) Increases the Distribution of the Glial-Derived Neurotrophic Factor (GDNF) Ligand Family in Rat Striatum and Enhances the Pharmacological Activity of Neurturin. *Exp. Neurol.* **2001,** *168* (1), 155–161.

Henkin, R. I. Intranasal Delivery to the Brain. *Nat. Biotechnol.* **2011,** *29* (6), 480.

Herce, H. D.; Garcia, A. E. Cell Penetrating Peptides: How Do They Do It? *J. Biol. Phys.* **2007,** *33* (5–6), 345–356.

Herve, F.; Ghinea, N.; Scherrmann, J. M. CNS Delivery Via Adsorptive Transcytosis. *AAPS J.* **2008,** *10* (3), 455–472.

Howard, M. A., 3rd; Gross, A.; Grady, M. S.; Langer, R. S.; Mathiowitz, E.; Winn, H. R.; Mayberg, M. R. Intracerebral Drug Delivery in Rats with Lesion-Induced Memory Deficits. *J. Neurosurg.* **1989,** *71* (1), 105–112.

Huang, R.; Ke, W.; Liu, Y.; Wu, D.; Feng, L.; Jiang, C.; Pei, Y. Gene Therapy Using Lactoferrin-Modified Nanoparticles in a Rotenone-Induced Chronic Parkinson Model. *J. Neurol. Sci.* **2010,** *290* (1–2), 123–130.

Imamura, A.; Uitti, R. J.; Wszolek, Z. K. Dopamine Agonist Therapy for Parkinson Disease and Pathological Gambling. *Parkinson. Relat. Disord.* **2006,** *12* (8), 506–508.

Isaac, M.; Holvey, C. Transdermal Patches: The Emerging Mode of Drug Delivery System in Psychiatry. *Ther. Adv. Psychopharmacol.* **2012,** *2* (6), 255–263.

Janson, C. G.; McPhee, S. W.; Leone, P.; Freese, A.; During, M. J. Viral-Based Gene Transfer to the Mammalian CNS for Functional Genomic Studies. *Trends Neurosci.* **2001,** *24* (12), 706–712.

Jellinger, K. A. Basic Mechanisms of Neurodegeneration: A Critical Update. *J. Cell. Mol. Med.* **2010,** *14* (3), 457–487.

Jiao, C. Y.; Delaroche, D.; Burlina, F.; Alves, I. D.; Chassaing, G.; Sagan, S. Translocation and Endocytosis for Cell-Penetrating Peptide Internalization. *J. Biol. Chem.* **2009,** *284* (49), 33957–33965.

Johnsen, K. B.; Burkhart, A.; Melander, F.; Kempen, P. J.; Vejlebo, J. B.; Siupka, P.; Nielsen, M. S.; Andresen, T. L.; Moos, T. Targeting Transferrin Receptors at the Blood-Brain Barrier Improves the Uptake of Immunoliposomes and Subsequent Cargo Transport into the Brain Parenchyma. *Sci. Rep.* **2017,** *7* (1), 10396.

Jollivet, C.; Aubert-Pouessel, A.; Clavreul, A.; Venier-Julienne, M. C.; Remy, S.; Montero-Menei, C. N.; Benoit, J. P.; Menei, P. Striatal Implantation of GDNF Releasing Biodegradable Microspheres Promotes Recovery of Motor Function in a Partial Model of Parkinson's Disease. *Biomaterials* **2004,** *25* (5), 933–942.

Jones, A. R.; Shusta, E. V. Blood–Brain Barrier Transport of Therapeutics via Receptor Mediation. *Pharm. Res.* **2007,** *24* (9), 1759–1771.

Kapp, S.; Gillespie-Lynch, K.; Sherman, L.; Hutman, T.; Klein, E.; Brown, T.; Sample, M.; Truitt, A.; Goergin, S.; Ortega, F. Targeted Delivery of Nerve Growth Factor Via Encapsulated Cell Biodelivery in Alzheimer Disease: A Technology Platform for Restorative Neurosurgery. *AJOB Neurosci.* **2017,** *8* (1), W1–W17.

Kelder, J.; Grootenhuis, P. D.; Bayada, D. M.; Delbressine, L. P.; Ploemen, J. P. Polar Molecular Surface as a Dominating Determinant for Oral Absorption and Brain Penetration of Drugs. *Pharm. Res.* **1999,** *16* (10), 1514–1519.

Kinch, M. S. An Overview of FDA-Approved Biologics Medicines. *Drug Discovery Today* **2015,** *20* (4), 393–398.

Kordower, J. H.; Charles, V.; Bayer, R.; Bartus, R. T.; Putney, S.; Walus, L. R.; Friden, P. M. Intravenous Administration of a Transferrin Receptor Antibody-Nerve Growth Factor Conjugate Prevents the Degeneration of Cholinergic Striatal Neurons in a Model of Huntington Disease. *Proc. Natl. Acad. Sci. U. S. Am.* **1994,** *91* (19), 9077–9080.

Kordower, J. H.; Emborg, M. E.; Bloch, J.; Ma, S. Y.; Chu, Y.; Leventhal, L.; McBride, J.; Chen, E. Y.; Palfi, S.; Roitberg, B. Z.; Brown, W. D.; Holden, J. E.; Pyzalski, R.; Taylor, M. D.; Carvey, P.; Ling, Z.; Trono, D.; Hantraye, P.; Deglon, N.; Aebischer, P. Neurodegeneration Prevented by Lentiviral Vector Delivery of GDNF in Primate Models of Parkinson's Disease. *Science* **2000,** 290 (5492), 767–773.

Kumar, P.; Wu, H.; McBride, J. L.; Jung, K. E.; Kim, M. H.; Davidson, B. L.; Lee, S. K.; Shankar, P.; Manjunath, N. Transvascular Delivery of Small Interfering RNA to the Central Nervous System. *Nature* **2007,** *448* (7149), 39–43.

Kumar, A.; Singh, A.; Ekavali, A Review on Alzheimer's Disease Pathophysiology and Its Management: An Update. *Pharmacol. Rep.* **2015,** *67* (2), 195–203.

Kurakhmaeva, K. B.; Djindjikhashvili, I. A.; Petrov, V. E.; Balabanyan, V. U.; Voronina, T. A.; Trofimov, S. S.; Kreuter, J.; Gelperina, S.; Begley, D.; Alyautdin, R. N. Brain Targeting of Nerve Growth Factor Using Poly(butyl cyanoacrylate) Nanoparticles. *J. Drug Target.* **2009,** *17* (8), 564–574.

Kusuhara, H.; Sugiyama, Y. Efflux Transport Systems for Drugs at the Blood-Brain Barrier and Blood-Cerebrospinal Fluid Barrier (Part 1). *Drug Discov. Today* **2001,** *6* (3), 150–156.

Lajoie, J. M.; Shusta, E. V. Targeting Receptor-Mediated Transport for Delivery of Biologics across the Blood-Brain Barrier. *Ann. Rev. Pharmacol. Toxicol.* **2015,** *55*, 613–631.

Lam, M. F.; Thomas, M. G.; Lind, C. R. Neurosurgical Convection-Enhanced Delivery of Treatments for Parkinson's Disease. *J. Clin. Neurosci.* **2011,** *18* (9), 1163–1167.

Larregina, A. T.; Morelli, A. E.; Spencer, L. A.; Logar, A. J.; Watkins, S. C.; Thomson, A. W.; Falo, L. D., Jr. Dermal-Resident CD14+ Cells Differentiate into Langerhans Cells. *Nat. Immunol.* **2001,** *2* (12), 1151–1158.

Lathuiliere, A.; Laversenne, V.; Astolfo, A.; Kopetzki, E.; Jacobsen, H.; Stampanoni, M.; Bohrmann, B.; Schneider, B. L.; Aebischer, P. A Subcutaneous Cellular Implant for Passive Immunization Against Amyloid-Beta Reduces Brain Amyloid and Tau Pathologies. *Brain J. Neurol.* **2016,** *139* (Pt. 5), 1587–1604.

Lewis, P. F.; Emerman, M. Passage through Mitosis Is Required for Oncoretroviruses But Not for the Human Immunodeficiency Virus. *J. Virol.* **1994,** *68* (1), 510–516.

Lin, C. Y.; Hsieh, H. Y.; Chen, C. M.; Wu, S. R.; Tsai, C. H.; Huang, C. Y.; Hua, M. Y.; Wei, K. C.; Yeh, C. K.; Liu, H. L. Non-Invasive, Neuron-Specific Gene Therapy by Focused Ultrasound-Induced Blood-Brain Barrier Opening in Parkinson's Disease Mouse Model. *J. Controlled Release* **2016,** *235*, 72–81.

Lin, C. Y.; Perche, F.; Ikegami, M.; Uchida, S.; Kataoka, K.; Itaka, K. Messenger RNA-Based Therapeutics for Brain Diseases: An Animal Study for Augmenting Clearance of Beta-Amyloid by Intracerebral Administration of Neprilysin mRNA Loaded in Polyplex Nanomicelles. *J. Controlled Release* **2016,** *235*, 268–275.

Loscher, W.; Potschka, H. Role of Drug Efflux Transporters in the Brain for Drug Disposition and Treatment of Brain Diseases. *Prog. Neurobiol.* **2005,** *76* (1), 22–76.

Lu, W. Adsorptive-Mediated Brain Delivery Systems. *Curr. Pharm. Biotechnol.* **2012,** *13* (12), 2340–2348.

Lu, C. T.; Zhao, Y. Z.; Wong, H. L.; Cai, J.; Peng, L.; Tian, X. Q. Current Approaches to Enhance CNS Delivery of Drugs Across the Brain Barriers. *Int. J. Nanomed.* **2014,** *9*, 2241–2257.

Lundberg, M.; Johansson, M. Is VP22 Nuclear Homing an Artifact? *Nat. Biotechnol.* **2001,** *19* (8), 713–714.

Mae, M.; Langel, U. Cell-Penetrating Peptides as Vectors for Peptide, Protein and Oligonucleotide Delivery. *Curr. Opin. Pharmacol.* **2006,** *6* (5), 509–514.

Maiti, P.; Manna, J.; Dunbar, G. L. Current Understanding of the Molecular Mechanisms in Parkinson's Disease: Targets for Potential Treatments. *Transl. Neurodegener.* **2017,** *6*, 28.

Matsuo, K.; Okamoto, H.; Kawai, Y.; Quan, Y. S.; Kamiyama, F.; Hirobe, S.; Okada, N.; Nakagawa, S. Vaccine Efficacy of Transcutaneous Immunization with Amyloid Beta Using a Dissolving Microneedle Array in a Mouse Model of Alzheimer's Disease. *J. Neuroimmunol.* **2014,** *266* (1–2), 1–11.

Mazarakis, N. D.; Azzouz, M.; Rohll, J. B.; Ellard, F. M.; Wilkes, F. J.; Olsen, A. L.; Carter, E. E.; Barber, R. D.; Baban, D. F.; Kingsman, S. M.; Kingsman, A. J.; O'Malley, K.; Mitrophanous, K. A. Rabies Virus Glycoprotein Pseudotyping of Lentiviral Vectors Enables Retrograde Axonal Transport and Access to the Nervous System after Peripheral Delivery. *Human Mol. Genetics* **2001,** *10* (19), 2109–2121.

McCarty, D. M.; DiRosario, J.; Gulaid, K.; Muenzer, J.; Fu, H. Mannitol-Facilitated CNS Entry of rAAV2 Vector Significantly Delayed the Neurological Disease Progression in MPS IIIB Mice. *Gene Ther.* **2009,** *16* (11), 1340–1352.

Menei, P.; Pean, J. M.; Nerriere-Daguin, V.; Jollivet, C.; Brachet, P.; Benoit, J. P. Intracerebral Implantation of NGF-Releasing Biodegradable Microspheres Protects Striatum against Excitotoxic Damage. *Exp. Neurol.* **2000,** *161* (1), 259–272.

Mercier, F.; Lefevre, G.; Huang, H. L.; Schmidli, H.; Amzal, B.; Appel-Dingemanse, S. Rivastigmine Exposure Provided by a Transdermal Patch Versus Capsules. *Curr. Med. Res. Opin.* **2007,** *23* (12), 3199–3204.

Mitragotri, S.; Burke, P. A.; Langer, R. Overcoming the Challenges in Administering Biopharmaceuticals: Formulation and Delivery Strategies. *Nat. Rev. Drug Discov.* **2014,** *13* (9), 655–672.

Mitrophanous, K.; Yoon, S.; Rohll, J.; Patil, D.; Wilkes, F.; Kim, V.; Kingsman, S.; Kingsman, A.; Mazarakis, N. Stable Gene Transfer to the Nervous System Using a Non-Primate Lentiviral Vector. *Gene Ther.* **1999,** *6* (11), 1808–1818.

Nguyen, T. T.; Giau, V. V.; Vo, T. K. Current Advances in Transdermal Delivery of Drugs for Alzheimer's Disease. *Indian J. Pharmacol.* **2017,** *49* (2), 145–154.

Nwogu, J. N.; Ma, Q.; Babalola, C. P.; Adedeji, W. A.; Morse, G. D.; Taiwo, B. Pharmacokinetic, Pharmacogenetic, and Other Factors Influencing CNS Penetration of Antiretrovirals. *AIDS Res. Treat.* **2016,** *2016*, 2587094.

Papazoglou, A.; Lundt, A.; Wormuth, C.; Ehninger, D.; Henseler, C.; Soos, J.; Broich, K.; Weiergraber, M. Non-restraining EEG Radiotelemetry: Epidural and Deep Intracerebral Stereotaxic EEG Electrode Placement. *J. Vis. Exp. Jove* **2016,** (112), 54216.

Pardeshi, C. V.; Belgamwar, V. S. Direct Nose to Brain Drug Delivery Via Integrated Nerve Pathways Bypassing the Blood-Brain Barrier: An Excellent Platform for Brain Targeting. *Expert Opin. Drug Deliv.* **2013,** *10* (7), 957–972.

Pfister, E. L.; DiNardo, N.; Mondo, E.; Borel, F.; Conroy, F.; Fraser, C.; Gernoux, G.; Han, X.; Hu, D.; Johnson, E.; Kennington, L.; Liu, P.; Reid, S. J.; Sapp, E.; Vodicka, P.; Kuchel, T.; Morton, A. J.; Howland, D.; Moser, R.; Sena-Esteves, M.; Gao, G.; Mueller, C.; DiFiglia, M.; Aronin, N. Artificial miRNAs Reduce Human Mutant Huntingtin Throughout the Striatum in a Transgenic Sheep Model of Huntington's Disease. *Human Gene Ther.* **2018,** *29* (6), 663–673.

Pillay, S.; Pillay, V.; Choonara, Y. E.; Naidoo, D.; Khan, R. A.; du Toit, L. C.; Ndesendo, V. M.; Modi, G.; Danckwerts, M. P.; Iyuke, S. E. Design, Biometric Simulation and Optimization of a Nano-Enabled Scaffold Device for Enhanced Delivery of Dopamine to the Brain. *Int. J. Pharm.* **2009,** *382* (1–2), 277–290.

Plenat, T.; Deshayes, S.; Boichot, S.; Milhiet, P. E.; Cole, R. B.; Heitz, F.; Le Grimellec, C. Interaction of Primary Amphipathic Cell-Penetrating Peptides with Phospholipid-Supported Monolayers. *Langmuir* **2004,** *20* (21), 9255–9261.

Poeschla, E. M.; Wong-Staal, F.; Looney, D. J. Efficient Transduction of Nondividing Human Cells by Feline Immunodeficiency Virus Lentiviral Vectors. *Nat. Med.* **1998,** 4 (3), 354–357.

Powell, E. M.; Sobarzo, M. R.; Saltzman, W. M. Controlled Release of Nerve Growth Factor from a Polymeric Implant. *Brain Res.* **1990,** *515* (1–2), 309–311.

Rabinowitz, J. E.; Samulski, J. Adeno-Associated Virus Expression Systems for Gene Transfer. *Curr. Opin. Biotechnol.* **1998,** 9 (5), 470–475.

Rankovic, Z. CNS Drug Design: Balancing Physicochemical Properties for Optimal Brain Exposure. *J. Med. Chem.* **2015,** *58* (6), 2584–2608.

Rathore, N.; Rajan, R. S. Current Perspectives on Stability of Protein Drug Products During Formulation, Fill and Finish Operations. *Biotechnol. Prog.* **2008,** *24* (3), 504–514.

Redmond, D. E., Jr.; McEntire, C. R.; Kingsbery, J. P.; Leranth, C.; Elsworth, J. D.; Bjugstad, K. B.; Roth, R. H.; Samulski, R. J.; Sladek, J. R., Jr. Comparison of Fetal Mesencephalic Grafts, AAV-Delivered GDNF, and Both Combined in an MPTP-Induced Nonhuman Primate Parkinson's Model. *Mol. Ther.* **2013,** *21* (12), 2160–2168.

Rodriguez Cruz, Y.; Strehaiano, M.; Rodriguez Obaya, T.; Garcia Rodriguez, J. C.; Maurice, T. An Intranasal Formulation of Erythropoietin (Neuro-EPO) Prevents Memory Deficits and Amyloid Toxicity in the APPSwe Transgenic Mouse Model of Alzheimer's Disease. *J. Alzheimer's Dis.: JAD* **2017,** *55* (1), 231–248.

Roe, T.; Reynolds, T. C.; Yu, G.; Brown, P. O. Integration of Murine Leukemia Virus DNA Depends on Mitosis. *EMBO J.* **1993,** *12* (5), 2099–2108.

Rose, J. M.; Audus, K. L. Receptor-Mediated Angiotensin II Transcytosis by Brain Microvessel Endothelial Cells. *Peptides* **1998,** *19* (6), 1023–1030.

Saito, Y.; Buciak, J.; Yang, J.; Pardridge, W. M. Vector-Mediated Delivery of ^{125}I-Labeled Beta-Amyloid Peptide A Beta 1-40 through the Blood-Brain Barrier and Binding to Alzheimer Disease Amyloid of the A beta 1-40/Vector Complex. *Proc. National Acad. Sci. United States of America* **1995,** *92* (22), 10227–10231.

Schenk, D.; Barbour, R.; Dunn, W.; Gordon, G.; Grajeda, H.; Guido, T.; Hu, K.; Huang, J.; Johnson-Wood, K.; Khan, K.; Kholodenko, D.; Lee, M.; Liao, Z.; Lieberburg, I.; Motter, R.; Mutter, L.; Soriano, F.; Shopp, G.; Vasquez, N.; Vandevert, C.; Walker, S.; Wogulis, M.; Yednock, T.; Games, D.; Seubert, P. Immunization with Amyloid-Beta Attenuates Alzheimer-Disease-Like Pathology in the PDAPP Mouse. *Nature* **1999,** *400* (6740), 173–177.

Shaltiel-Karyo, R.; Frenkel-Pinter, M.; Egoz-Matia, N.; Frydman-Marom, A.; Shalev, D. E.; Segal, D.; Gazit, E. Inhibiting Alpha-Synuclein Oligomerization by Stable Cell-Penetrating Beta-Synuclein Fragments Recovers Phenotype of Parkinson's Disease Model Flies. *PLoS One* **2010,** *5* (11), e13863.

Singer, O.; Marr, R. A.; Rockenstein, E.; Crews, L.; Coufal, N. G.; Gage, F. H.; Verma, I. M.; Masliah, E. Targeting BACE1 with siRNAs Ameliorates Alzheimer Disease Neuropathology in a Transgenic Model. *Nat. Neurosci.* **2005,** *8* (10), 1343–1349.

Snyder, R. O. Adeno-Associated Virus-Mediated Gene Delivery. *J. Gene Med.* **1999,** *1* (3), 166–175.

Soderquist, R. G.; Mahoney, M. J. Central Nervous System Delivery of Large Molecules: Challenges and New Frontiers for Intrathecally Administered Therapeutics. *Expert Opin. Drug Deliv.* **2010,** 7 (3), 285–293.

Song, B. W.; Vinters, H. V.; Wu, D.; Pardridge, W. M. Enhanced Neuroprotective Effects of Basic Fibroblast Growth Factor in Regional Brain Ischemia After Conjugation to a Blood-Brain Barrier Delivery Vector. *J. Pharmacol. Exp. Ther.* **2002,** *301* (2), 605–610.

Stiles, D. K.; Zhang, Z.; Ge, P.; Nelson, B.; Grondin, R.; Ai, Y.; Hardy, P.; Nelson, P. T.; Guzaev, A. P.; Butt, M. T.; Charisse, K.; Kosovrasti, V.; Tchangov, L.; Meys, M.; Maier, M.; Nechev, L.; Manoharan, M.; Kaemmerer, W. F.; Gwost, D.; Stewart, G. R.; Gash, D. M.; Sah, D. W. Widespread Suppression of Huntingtin with Convection-Enhanced Delivery of siRNA. *Exp. Neurol.* **2012,** *233* (1), 463–471.

Tan, C. C.; Yu, J. T.; Wang, H. F.; Tan, M. S.; Meng, X. F.; Wang, C.; Jiang, T.; Zhu, X. C.; Tan, L. Efficacy and Safety of Donepezil, Galantamine, Rivastigmine, and Memantine for the Treatment of Alzheimer's Disease: A Systematic Review and Meta-Analysis. *J. Alzheimer's Dis.: JAD* **2014,** *41* (2), 615–631.

Tang, W. L.; Zhao, H. Industrial Biotechnology: Tools and Applications. *Biotechnol. J.* **2009,** *4* (12), 1725–1739.

Tereshchenko, J.; Maddalena, A.; Bahr, M.; Kugler, S. Pharmacologically Controlled, Discontinuous GDNF Gene Therapy Restores Motor Function in a Rat Model of Parkinson's Disease. *Neurobiol. Dis.* **2014,** *65*, 35–42.

Thorne, R. G.; Pronk, G. J.; Padmanabhan, V.; Frey, W. H., 2nd. Delivery of Insulin-Like Growth Factor-I to the Rat Brain and Spinal Cord along Olfactory and Trigeminal Pathways Following Intranasal Administration. *Neuroscience* **2004,** *127* (2), 481–496.

Tietz, S.; Engelhardt, B. Brain Barriers: Crosstalk between Complex Tight Junctions and Adherens Junctions. *J. Cell Biol.* **2015,** *209* (4), 493–506.

Ulbrich, K.; Hekmatara, T.; Herbert, E.; Kreuter, J. Transferrin- and Transferrin-Receptor-Antibody-Modified Nanoparticles Enable Drug Delivery across the Blood-Brain Barrier (BBB). *Eur. J. Pharm. Biopharm.* **2009,** *71* (2), 251–256.

Ulbrich, K.; Knobloch, T.; Kreuter, J. Targeting the Insulin Receptor: Nanoparticles for Drug Delivery across the Blood-Brain Barrier (BBB). *J. Drug Targeting* **2011,** *19* (2), 125–132.

van Es, M. A.; Hardiman, O.; Chio, A.; Al-Chalabi, A.; Pasterkamp, R. J.; Veldink, J. H.; van den Berg, L. H. Amyotrophic Lateral Sclerosis. *Lancet* **2017,** *390* (10107), 2084–2098.

Van Vliet, K. M.; Blouin, V.; Brument, N.; Agbandje-McKenna, M.; Snyder, R. O. The Role of the Adeno-Associated Virus Capsid in Gene Transfer. *Methods Mol. Biol.* **2008,** *437*, 51–91.

Wang, C.; Wang, C. M.; Clark, K. R.; Sferra, T. J. Recombinant AAV Serotype 1 Transduction Efficiency and Tropism in the Murine Brain. *Gene Ther.* **2003,** *10* (17), 1528–1534.

Wang, Y. Y.; Lui, P. C.; Li, J. Y. Receptor-Mediated Therapeutic Transport across the Blood-Brain Barrier. *Immunotherapy* **2009,** *1* (6), 983–993.

Wei, Y. J.; Stuart, B.; Zuckerman, I. H. Use of Anti-Parkinson Medications among Elderly Medicare Beneficiaries with Parkinson's Disease. *Am. J. Geriatr. Pharmacother.* **2010,** *8* (4), 384–394.

Weiss, N.; Miller, F.; Cazaubon, S.; Couraud, P. O. The Blood-Brain Barrier in Brain Homeostasis and Neurological Diseases. *Biochim. Biophys. Acta* **2009,** *1788* (4), 842–857.

Wen, Z.; Yan, Z.; Hu, K.; Pang, Z.; Cheng, X.; Guo, L.; Zhang, Q.; Jiang, X.; Fang, L.; Lai, R. Odorranalectin-Conjugated Nanoparticles: Preparation, Brain Delivery and Pharmacodynamic Study on Parkinson's Disease Following Intranasal Administration. *J. Controlled Release* **2011,** *151* (2), 131–138.

Woodard, K. T.; Liang, K. J.; Bennett, W. C.; Samulski, R. J. Heparan Sulfate Binding Promotes Accumulation of Intravitreally Delivered Adeno-associated Viral Vectors at the Retina for Enhanced Transduction but Weakly Influences Tropism. *J. Virol.* **2016,** *90* (21), 9878–9888.

Wu, D.; Yang, J.; Pardridge, W. M. Drug Targeting of a Peptide Radiopharmaceutical through the Primate Blood-Brain Barrier *In Vivo* with a Monoclonal Antibody to the Human Insulin Receptor. *J. Clin. Invest.* **1997,** *100* (7), 1804–1812.

Xiao, G.; Gan, L. S. Receptor-Mediated Endocytosis and Brain Delivery of Therapeutic Biologics. *Int. J. Cell Biol.* **2013,** *2013*, 703545.

Xue, Y. Q.; Ma, B. F.; Zhao, L. R.; Tatom, J. B.; Li, B.; Jiang, L. X.; Klein, R. L.; Duan, W. M. AAV9-Mediated Erythropoietin Gene Delivery into the Brain Protects Nigral Dopaminergic Neurons in a Rat Model of Parkinson's Disease. *Gene Ther.* **2010,** *17* (1), 83–94.

Yan, L.; Wang, H.; Jiang, Y.; Liu, J.; Wang, Z.; Yang, Y.; Huang, S.; Huang, Y. Cell-Penetrating Peptide-Modified PLGA Nanoparticles for Enhanced Nose-To-Brain Macromolecular Delivery. *Macromol. Res.* **2013,** *21* (4), 435–441.

Youn, P.; Chen, Y.; Furgeson, D. Y. A Myristoylated Cell-Penetrating Peptide Bearing a Transferrin Receptor-Targeting Sequence for Neuro-Targeted siRNA Delivery. *Mol. Pharm.* **2014,** *11* (2), 486–495.

Zhang, C.; Chen, J.; Feng, C.; Shao, X.; Liu, Q.; Zhang, Q.; Pang, Z.; Jiang, X. Intranasal Nanoparticles of Basic Fibroblast Growth Factor for Brain Delivery to Treat Alzheimer's Disease. *Int. J. Pharm.* **2014,** *461* (1–2), 192–202.

Zlokovic, B. V. The Blood-Brain Barrier in Health and Chronic Neurodegenerative Disorders. *Neuron* **2008,** *57* (2), 178–201.

CHAPTER 4

Genes in Genetic Disease

SUKANYA BHOUMIK and SYED IBRAHIM RIZVI*

Department of Biochemistry, University of Allahabad, Allahabad, India

Corresponding author. E-mail: sirizvi@gmail.com

ABSTRACT

Genes are the hereditary factors that are passed from generation to generation and are responsible for determining the genotypic as well as phenotypic traits in an individual. Every human being has about 20,000–25,000 genes which encodes for a variety of polypeptides and proteins. The transfer of the genetic information contained in a gene produces messenger RNA (mRNA) which is further translated into proteins. The information in the DNA is sometimes tweaked due to certain mutations in the gene. These mutations are the cause of a number of genetic diseases. Gene mutations can develop either as spontaneous (without any external agent but may need time to express in the genome) or induced mutations (incorporated because of a mutagen which may either be an external radiative agent or a biological agent). Broadly, gene mutations are classified in three categories as single gene mutation, multifactorial inheritance disorder, and chromosomal disorders. Each human possibly has a few gene mutations in their cells, but the expression depends whether they are dominant or recessive. A large number of genetic disorders are known that arise from either type of gene mutations and are almost lethal or may cause variety of abnormalities. Disease susceptibility of an individual depends on its genomic confirmation as well as external exposure.

4.1 INTRODUCTION

The genome comprises the entire genetic composition of our body including the coding as well as noncoding regions plus the information contained in the mitochondrial DNA and chloroplasts. More than 3 million DNA base pairs make up the human genome and this information is contained in each and every nucleated cell of our body. This hereditary information gets translated into proteins which constitutes the structural framework and is necessary to perform the metabolic functions of the body. The genome is divided into smaller units, known as gene, which have the ability to specifically encode for a single protein or polypeptide. The human genome comprises of at least 20,000–25,000 genes, as estimated by the Human genome project, which are functionally active, the exons. The rest of the genome is the noncoding part, termed as introns which may be silenced, or may be translated into nonfunctional RNA which is removed by RNA splicing. Each gene codes for a specific protein. The genes are inherited to us from our parents, thus they are in sets of two; one from the mother and another from the father. The DNA is found in a coiled-coil conformation inside the nucleus and is known as chromosome supported by histone proteins. In each cell, there are sets of 22 pairs of chromosome which are the same in all humans, known as autosomes, and the 23rd pair is the sex chromosome, comprising XX in females and XY in males.

4.1.1 MUTATIONS CAUSE GENETIC VARIATION

The position or location of a particular gene on the chromosome is known as its genetic locus. The arrangement of genes on the chromosome and the linkage between the genes on adjacent chromosomes involves the genetic linkage. The genome is an unstable entity. This genomic instability is due to different kinds of heritable and inheritable changes. These changes can occur suddenly at any time point of life which interrupts the normal coding of the gene. These alterations in the genomic sequence change the information carried by the particular gene segment. These changes causing misplaced nucleotide(s) which alters the genotype is called mutation. Although mutations are a randomly occurring event and are mostly deleterious but they are also necessary for evolution and to adapt to the environmental factors—both physical and biological.

Mutational changes can occur either in the autosomes (somatic cells) or sex chromosomes. These alterations are not always harmful since they provide a small source of genetic variability which is absolutely essential. A large number of genetic disorders are known due to the different types of genetic alterations such as chromosomal aberrations, changes in individual nucleotides or ploidy.

4.1.2 BASIS OF GENE MUTATION

Gene mutations can be caused by two ways: either spontaneous or induced. Spontaneous mutations occur naturally and do not need any external agent. It can occur due to changes/displacement/absence of nucleotides. These changes can occur either during DNA replication, lesions in DNA, or transposition in gene caused by transposable elements during normal cell cycle. Induced mutations arise because of a causative agent generally a mutagen which causes a change in the original DNA. Induced mutations can be caused by radiation damage to DNA such as UV ray, X ray, or β ray or by biological causative agents such a transposons, chemical mutagens, and intercalating agents. Since spontaneous mutations are naturally occurring they take a long time to get incorporated into the genome whereas induced mutations occur at a much greater rate because of their instant reactivity with the parent DNA. Gene mutations interfere with the normal process of DNA replication or transcription or protein synthesis machinery, and carry over the defect to the next step of the central dogma.

4.1.3 GENE MUTATIONS AND THEIR TYPES

Genetic diseases are caused due to changes in the sequence of the genome causing an interruption such as addition, deletion, or displacement of a nucleotide in the normal gene sequence. All diseases arise from a disturbed genetic component. Some are present from birth and are transmitted to them from their parent, known as sex-linked (X-linked) inheritance while other are acquired during the person's lifetime, known as autosomal inheritance. Some diseases are passed to generations by transmission of the gene carrying it. Genetic disorders are grouped into the categories described below:

1. **Single gene mutation (monogenetic disorder/substitution mutation):** Is caused by mutation in a single gene, either on one chromosome or both the chromosomes. The chances of occurrence of this type of mutation depend on whether the disease causing gene is dominant or recessive. Diseases carrying dominant genes occur when the causative gene is present on even a single chromosome such as Huntington's disease and Marfan syndrome whereas recessive diseases are caused when the causative gene is present on both the chromosome such as sickle cell anemia, cystic fibrosis, and Tay–Sachs disease.
2. **Multifactorial inheritance disorder:** Is caused not only due to change in the normal pattern of gene/(s) expression but also involves external factors such as our lifestyle including eating patterns, exposure to environmental toxins and chemicals, etc. Disorders of this kind cannot occur solely only on the basis of gene mutation without any external factor affecting it, although the presence of the gene mutation greatly increases the chances of any such disorder. Multifactorial genetic conditions have a chance of passing down to family but accurate chances of it cannot be known because these diseases do not show any specific pattern of inheritance. Some traits such as hair color and height are a result of the interaction between different genes and the environment. Examples of this kind of disorder include many conditions such as obesity, diabetes, different forms of cancer, cardiovascular conditions, hypertension and cholesterol level, cleft lip, respiratory problems such as asthma and allergy.
3. **Chromosomal disorders:** Human beings contain 23 pairs of chromosome, 22 pairs of autosomes, and a pair of sex chromosome. Females contain XX as the sex chromosome while males have XY as the sex chromosome. Whether an X or Y chromosome is received from the sperm cell determines the sex of the child. Therefore, the female is represented as 46,XX and male as 46,XY. Chromosomal aberrations are caused by any change in the normal organization of a single chromosome or a pair of chromosome (Tartaglia et al., 2013). Chromosomal changes may be inherited or spontaneous. The various types of chromosomal changes are described below:

   a) **Ploidy (alteration in the number of chromosome in cells)** Every individual receives a pair of chromosomes, one from each parent. Possibilities of error formation always remain especially during the separation of chromosomes in DNA replication, or an inherited error received from the parent during gamete formation. A change in the number of sets of chromosome is called **euploidy.** Whereas **aneuploidy** involves addition or deletion of a single chromosome, not in a whole set. Diseases of this type include Down's syndrome, Edward's syndrome, Turner syndrome, Klienfelter syndrome, and many more.

b) **Chromosomal aberrations (morphological changes in chromosome)**

Deletion: It involves loss of a part of a chromosome either from any end of the chromosome or from any middle segment, termed as terminal and interstitial deletion, respectively. When the lost part contains any vital genetic information, it leads to a disturbed condition known as genetic imbalance. An example of deletion is the 5p-syndrome in which, a part of the short arm of chromosome 5 is deleted, leading to health disturbances.

Duplication: It results in an additional extra copy of any part of the chromosome which can be copied in the same sequence known as tandem duplication or copied in the opposite order known as reverse form.

Inversions: Any disturbance which causes two breaks at any region of a chromosome and lets the chromosome rejoin after a 180° twist is known as inversion. Thus, inversions do not change the amount of genetic material and so does not show much phenotypic change. If the inverted segment contains an essential genetic information, then such inversions can be lethal.

Translocations: It involves any interchange of parts between two nonhomologous chromosomes creating two new translocation chromosomes. They can be of two types: reciprocal and nonreciprocal (Robertsonian) translocations.

Reciprocal translocation: In this type, there is a mutual exchange of segments between two nonhomologous chromosomes. This generally happens when any part from both the chromosome pairs breaks at any point during the egg or sperm formation. Overall, the amount of genetic material remains the same and there is no gain or loss of chromosome.

Robertsonian translocation: Here, there is no mutual exchange of chromosome parts but transfer of a single broken chromosome segment to another chromosome from a different pair.

Epigenetics: Not always are the observed phenotypes because of the change in the genetic constitution. Any slight modification in the DNA sequence such as methylation, phosphorylation, acetylation, ubiquitylation, and sumoylation without any change in the original sequence of genes is termed epigenetics. These changes sometimes determine the active functioning of the gene, the proteins translated, thus influencing the phenotype (Weinhold, 2006).

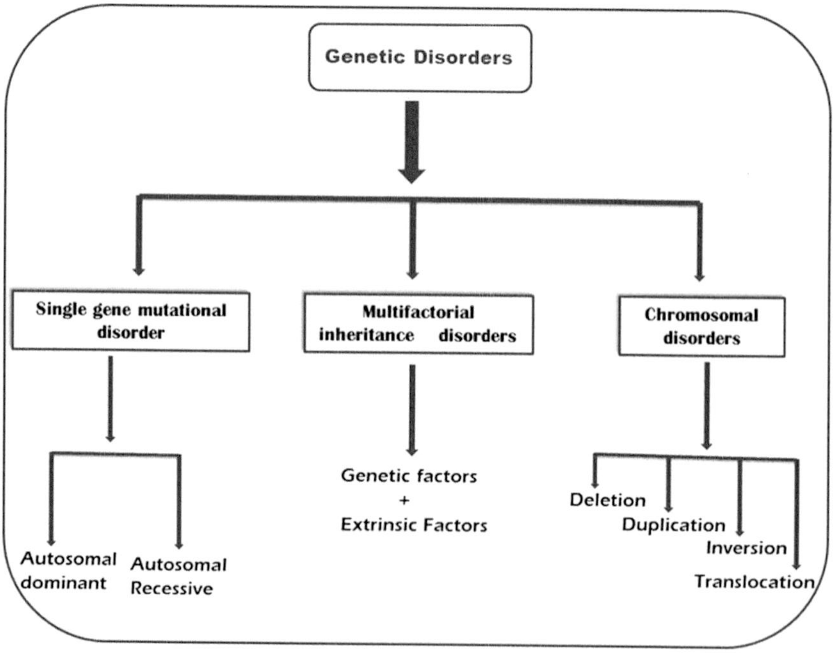

FIGURE 4.1 Classification of genetic mutation.

4.1.3.1 *AUTOSOMAL DOMINANT DISEASES*

Huntington's disease: This disease is caused due to a mutation in the HTT gene which synthesizes a protein named huntingtin. The DNA segment encoding the disease causing portion of the gene are CAG repeat units. In any normal individual, the gene segment is repeated 10–35 times but in the diseased individual it repeats more than 37 times. The Huntington disease gene is known as *IT-15* (interesting transcript number 15) which consists of a trinucleotide repeat unit CAG whose proteins denote polyglutamine units. This is an autosomal dominant disorder as the presence of a single copy is sufficient to cause the disease, either from any parent. The parent protein in which the trinucleotide CAG unit resides also decides the neuropathology of the disease. The accumulation of the mutated huntingtin protein in the brain determines the intensity of the disease. Such as in adult onset type of this disease, the mutated region, that is, the N-terminal part is present in the neuritis, whereas the cortex and striatum is the part of accumulation in case of juvenile onset type (Aronin et al., 1999).

The huntingtin protein weighing 350 kDa starts to appear early from the beginning of the development of brain. Ultrastructural studies of the huntingtin protein have shown its association with the Golgi vesicles, clathrin coated pits, and vesicles for transport. Interaction of the huntingtin protein with the cytoplasmic proteins such as huntingtin associated protein 1 (HAP-1), huntingtin interacting protein (HIP-1), and calmodulin all eligible for movement along microtubules and vesicular transport (Zheng and Diamond, 2012).

Marfan syndrome: It is a problem of the connective tissue, which supports the structural framework of the body tissues and organs. This syndrome is caused due to defect in the gene encoding the protein fibrillin-1 (*FBN1*) whose presence is found in the microfibrils of the extracellular matrix. Structurally a glycoprotein, especially rich in cysteine, fibrillin, weighing 350 kDa, was mapped on the long arm of chromosome 15. This protein in turn, increases synthesis of transforming growth factor beta (TGF-β). The TGF-β is stored in the microfibrils itself and which in turn acts as a switch regulating the expression of TGF-β. TGF-β when released by the microfibrils are activated and inactive when present within the microfibrils (Robinson and Godfrey, 2000).

4.1.3.2 AUTOSOMAL RECESSIVE DISEASES

Sickle cell anemia: It is a disease of the red blood cells which is caused due to a defect in the hemoglobin protein which has the function of transporting oxygen in the erythrocytes. Hemoglobin molecule consists of 2 α chains and 2 β chains. Hemoglobin S is the causative factor of sickle cell anemia and is caused when two copies of Hemoglobin S are inherited from both the parents. The sickle cell hemoglobin (HbS) consists of a glutamate residue in place of a valine normally present in the HbA in the β chain. This difference causes changes in the quaternary structure of hemoglobin which results in the aggregation of deoxyHbS into sickle shaped fibers. These fibers cause the RBCs shape to be distorted which blocks the flow of blood to every cells of the body causing anemic conditions, puffiness of the hands and feet, fatigue and irritation, etc. In case a single distorted gene is inherited from any parent, then the condition is known as sickle cell trait (Sergeant, 2013).

Cystic fibrosis: This disease is due to the blockage caused in the passage ways such as the respiratory canal, alimentary canal, reproductive system due to disbalance in the movement of salts in and out of the cells. The causative agent is mucus that blocks these passageways functionally which is enrolled to protect and border these linings. In case of cystic fibrosis, the mucus becomes thick and more sticky and an imbalance occurs in the concentration of salts in sweat.

The chromosomal location of the gene causing cystic fibrosis is at the long arm (q) of chromosome 7, positioned at 31.2. Cystic fibrosis trans-membrane conductance regulator (CFTR) gene synthesizes the protein cystic fibrosis transmembrane conductance regulator which works in the membranous compartment of cells specific for sweat, mucus, tears, saliva, and digestive secretions. Ions involved in the CFTR transport are chloride ions moving in both directions and the sodium ion moves out of the cell. The production of mucus is controlled by the water transport into tissues by chloride ions. Mutations in the amino acids can occur in a number of ways in cystic fibrosis. Deletion in the 508th amino acid is known as delta F508 is the most frequently occurring mutation out the several types of mutations in the CFTR gene. The mutations alter the stability or functioning of the chloride ion channels which disturbs the water movement and chloride transport in and out of cells. These changes lead to the production of extremely thick mucus blocking the passageways of lungs and pancreas causing obstruction in the air channels and glands (Dodge, 2015; Davies, 2014).

Tay–Sachs disease: This is a disorder of the neurons involving the brain and the spinal cord. Symptoms of this are more prominent in infants than in adults or adolescents. The causative gene of this disease is HEXA forming the alpha subunit which together with the HEXB gene constituting the beta subunit forms the fully functional enzyme, beta-hexosaminidase A. This enzyme is present in the lysosomes and mutation in the HEXA gene leads to formation of ganglioside GM2 leading to a condition GM gangliosidoses, a disorder of the neurons. Due to the non-breakdown of fatty acid derivatives known as gangliosides they accumulate in the lysosomes disturbing the functioning of the neurons (Tsuji, 2013.)

The cytogenic location of HEXA gene is the long arm of chromosome 15 at position 23.

4.1.4 SOME OF THE MULTIFACTORIAL DISORDERS ARE DESCRIBED BELOW

Obesity: The problem of obesity has drastically risen in the last two decades due to increased consumption of caloric rich food particularly in the United States and other western countries. Obesity is the condition of increased accumulation of fats in the body which leads to its interference with other diseases such as type 2 diabetes, hypertension, atherosclerosis, cancer, and other degenerative disorders. Although environmental and lifestyle behavior play a significant role in the etiology of obesity, the genetic component of obesity has also been detected using twins as models and the identified types are monogenic or syndromic obesity and polygenic obesity. The mouse *ob* gene was the cloned animal model for obesity. Its homologue, identified in humans is the leptin gene weighing 16-kDa stored in adipocytes and its receptor located in the hypothalamus. The functional role of leptin is to maintain our appetite, body weight, and energy intake. Leptin is the gene recognized in humans for regulation of obesity and related traits. The identification of the leptin-melanocortin pathways introduced many genes such as the pro-opiomelanocortin (POMC) gene associated positively with the leptin pathway (Xia and Grant, 2013). α-melanocortin stimulating hormone receptor (MC4R) and prohormone convertase-1 are the genes responsible for obesity. Besides these genes because of the multifactorial nature of obesity there are a number of other genes which are also closely associated with other disorders. The glutamic acid decarboxylase gene (GAD2), Pre-B cell colony enhancing factor (PBEF1), recently reported as visfatin, and the growth hormone secretagogue receptor (GHSR) and, the receptor for ghrelin are involved in appetite regulation and feeding (Walley et al., 2006).

Cardiovascular disease: Coronary heart diseases involve a number of disorders such as myocardial infarction, vascular disorders of heart, congenital heart diseases, etc. and these groups of disorders comprise the highest number of deaths in western nations (Poulter, 1999). Cardiovascular diseases are caused due to external physical, biological, or chemical factors and internal factors which are heritable. Research studies suggest either the involvement of single gene because of the simple trends seen in pattern of inheritance or the complex of multiple genes and also interplay of other external factors (Kathiresan and Srivastava, 2013). A leading risk factor for cardiovascular disease is obesity. Avoiding high caloric diets and doing regular exercise can to some extent reduce the risk. Also hypertension, cigarette smoke, high cholesterol levels are associated risk factors. Increasing urbanization, industrialization, and globalization are additional factors showing increased tendency to cause heart diseases.

Breast cancer: There is seen an increasingly higher incidence of breast cancer presently causing a major concern. Breast cancer is caused due to a major interaction between the genes and the environment. It is known that breast cancer is a hereditary disease. Some genes responsible for this disease have been identified such as the BRCA1 and BRCA2. Exposure to the hormone estrogen is a cause for this disease with its chances increasing on increased exposure. Obesity has been linked to breast cancer because obese women are at high chances of exposure to estrogen. Dietary factors such as consumption of meat, alcohol, high-fat foods are associated with an increased susceptibility to causing breast cancer.

The BRCA1 genomic entity, structurally a zinc finger domain at the amino terminus is about 100 kilobases (kb). Evidence suggest that the BRCA1 gene is an autosomal dominant gene and germinal mutations in the BRCA1 gene in breast or ovarian cancer patients is seen in almost 80% of cases. The BRCA1 gene is present in chromosome 17q and is known to be a tumor suppressor gene, that is, only if both alleles of the gene are lost that the disease can be malignant. The BRCA2 cDNA sequence comprises about 70 kb of genomic DNA. It is also an autosomal dominant gene known to cause 65-80% of breast cancer cases (Martin and Weber, 2000).

Diabetes mellitus: Type 1 diabetes mellitus also known as insulin dependent diabetes mellitus is a known cause of morbidity worldwide and a multifactorial disease. Since the type 1 diabetes mellitus is an autoimmune condition so the genetic loci of this disease are on chromosome 6p21 of the major histocompatibility complex (MHC) (Cordell and Todd, 1995).

The type 2 diabetes mellitus is also a multifactorial disease with external factors such as obesity, diet, age involved. In case of absence of exposure to external factors, the genetic determinants present may remain silent. Some of the gene products which influence the insulin secretion pathway are the glucagon receptor, the peroxisome proliferator activation-α, insulin receptor substrate-1 (IRS-1) (Wijmenga et al., 2001).

4.1.5 THE VARIOUS CHROMOSOMAL ABNORMALITIES INVOLVING BOTH STRUCTURAL AND NUMERICAL VARIATIONS ARE LISTED AS FOLLOWS

Down syndrome: Also known as trisomy 21, Down syndrome causes appearance of some typical facial characteristics, heart defect since birth, less intellectual ability. Trisomy 21 results from an extra copy of chromosome 21 in all cells of the body. This is a case of translocation where some part of the 21st chromosome translocates or gets attached to another chromosome during gamete formation in an individual parent or after fusion in a fetus. The extra copy of chromosome 21 results in three copies of chromosome causing the conditions related to Down syndrome. There is another rare case (5%) of Down syndrome children showing Robertsonian translocation where the long arm of chromosome 21 fuses with the long arm of chromosome 14. This type of inheritance is generally passed onto generations by carriers who are unaffected. Occurrence of Down syndrome in any generation is due to meiotic division which results in the child carrying three copies of chromosome 21 (Asim et al., 2015).

Edward syndrome: Similar to trisomy 21, Edward syndrome is a case of trisomy 18, that is, three copies of chromosome 18. This is a rare condition but can be fatal for the baby as it is difficult for the baby to survive for more than a year with this condition. In case of mosaic trisomy 18, only some cells of the body have an extra copy of chromosome 18, the severity of which depends on the number of affected cells (Haldeman-Englert, 2018). The long arm (q) of chromosome 18 translocates or attaches to a different chromosome pair during gamete formation, as a result of disjunction. If the gametic cell is passed on to the child, then all the cells of the body will have an extra chromosome 18 (Satge et al., 2016).

Patau syndrome: Also known as trisomy 13, Patau syndrome occurs because of an extra copy of chromosome 13 in all cells of the body. Due to an error in cell division during the separation of gametes, an extra copy of the 13th chromosome gets translocated while forming the gametes (egg or sperm) or early in embryonic development (Tartaglia et al., 2013).

Cri-du-chat syndrome: This syndrome represents a condition of deletion where a part of the short arm of chromosome 5 is deleted. This condition is also known as the cat's cry or the 5p minus syndrome. The severity of the disease depends on the portion of the chromosome deleted. The deleted part of the chromosome also contains a gene named CTNND 2 (catenin delta 2). This gene synthesizes the protein delta-catenin. Delta-catenin serves a very important role in the brain where they help relay information between synapses. They have cell adhesion properties, therefore help cells remain at their place, directs the nerve cells to their proper location. Overall, delta catenin is a neuronal regulator, is present in the dendrites and involved in functioning of the synapses. It is also involved in the Wnt signaling pathway, interacts with beta-catenin, and maintains overall integrity of neurons (Lu et al., 2016).

Wolf-Hirschhorn syndrome: It is a severe condition involving deletion of the part of the short arm of chromosome 4, often denoted as 4p-. The multiple and varied signs of this syndrome can be attributed to the genes lost in the deleted portion of the chromosome. NSD2, LETM1, and MSX1 are the genes lost in the deleted part. Loss of nuclear receptor binding SET domain protein 2 (NSD2) is responsible for the distinct facial features such as broad forehead, asymmetrical face, and a small head. The leucine zipper and EF-hand containing transmembrane protein 1 (LETM1) protein is associated with the electrical signals in the brain mitochondrial cells. It is also involved in transfer of calcium ions in mitochondria. The msh homeobox 1 (MSX1) gene, part of a homeobox gene regulates the early development of many body structures mainly the teeth and the nails.

Wolf-Hirschhorn syndrome occurs as a random event during the formation of eggs or sperm. Otherwise formation of a ring chromosome can also be a cause in which the chromosome breaks at two points and the remaining ends join to form a ring. Or the deleted chromosome 4 can be inherited from a parent who has a rearranged chromosome 4 with another chromosome (Battaglia et al., 2015)

Jacobsen syndrome: This is caused due to deletion at the end of the long arm of chromosome 11. Approximately, 5–16 Mb of genetic material is lost in the deleted part which involves the functioning of 170–340 genes. Although Jacobsen syndrome is mostly caused as a chance event during egg or sperm formation or during embryonic development, in some cases, balanced translocation occurs in which the deleted chromosomal 11 segment is exchanged with a segment of another chromosome. Here no part of the genetic material is gained or lost (Grossfeld et al., 2015).

Klinefelter's syndrome: It is a chromosomal abnormality affecting the males caused because of the sex chromosome. Every normal male has a set of XY chromosome and every normal female has an XX set but in this condition, the male has an extra X chromosome. Thus, Klinefelter males have 47, XXY as their sex chromosome. These males do not have proper sexual maturity because of reduced levels of testosterone. Their testes are incompletely developed, have female characteristics such as enlarged breasts, less hairy bodied, and are generally impotent. They are prone to develop an inflammatory disease systemic lupus erythematosus. Sometimes individuals with Klinefelter syndrome may have even more than an extra X chromosome such as 48, XXXY or 49, XXXXY. They show symptoms worse than their counterparts having normal Klinefelter syndrome 47, XXY, showing more the number of sex chromosome, more the deformity. In some cases, a person might not have an extra X chromosome in each cell such as (46, XY and 47, XXY) as the chromosomal constitution. This condition is known as chromosomal mosaicism. Diagnosis of the XXY genetic makeup is generally identifiable by karyotyping because of the mostly same signs of both XY and XXY. The short stature-homeobox-containing gene (SHOX) present on the pseudoautosomal region 1 of the short arm p of chromosome X influences the extra growth in Klinefelter's individuals. The androgen receptor, activated by binding of the hormone testosterone and other androgenic hormones contains the CAG repeat units in Klinefelter's syndrome patients (Groth et al., 2013).

Turner syndrome: This chromosomal aneuploidy occurring in females is because of the presence of only a single X chromosome in place of XX normally present in females. The other X chromosome may be missing or structurally deformed. This is an example of monosomy because only single X is present in each cell or some of the cells only as in mosaicism. The other defective X chromosome may have a missing short arm p (Xp) or absent long arm q (Xq) may lead to infertility problems and irregular periods. The SHOX gene is responsible for the shortened height and skeletal deformities in females with turner syndrome. Also, the SHOX gene is known regulate growth and bone development (Levitsky et al., 2015). The phenotype of females with turner syndrome may vary depending on the lost part of X chromosome or totally missing X chromosome. Mosaicism, imprinting, X chromosome inactivation, or gene dosage effects are the factors for the phenotypic diversity seen in turner syndrome (Tartaglia et al., 2013).

4.2 CONCLUSION

The variety of human diseases caused due to factors mediated through genetic abberations are varied and pleiotropic in nature. Genetic variation within population or families is passed on to generations and their expressivity depends either on their dominance, or their joint effect with other genes involved or under the influence of environmental factors. The genetic susceptibility of an individual to a disease depends on his/her family's exposure to the genomic and environmental conditions. The diseases inherited with a Mendelian pattern of inheritance show a phenomenon called allelic heterogeneity where a disease is caused due to multiple mutations on the same gene (Hernandez and Blazer, 2006).

KEYWORDS

- **genes**
- **mutations**
- **chromosomal disorders**
- **syndrome**
- **genomic**

REFERENCES

Aronin, N.; Kim, M.; Laforet, G.; DiFiglia, M. Are there Multiple Pathways in the Pathogenesis of Huntington's disease? *Phil. Trans. R. Soc. Lond. B.* **1999,** 354–1003.

Asim, A., Kumar, A., Muthuswamy, S., Jain, S., Agarwal, S., Down Syndrome: An Insight of the Disease. *J. Bio. Med. Sci.* **2015,** *22*, 41.

Battaglia, A.; Carey, J. C.; South, S. T.; Wolf-Hirschhorn Syndrome: A Review and Update. *Am. J. Med. Genet.* **2015,** *169C*, 216–223.
Cordell, H. J.; Todd, J. A. Multifactorial Inheritance in Type 1 Diabetes. *Trends Genet.* **1995,** *11*(12).
Davies, J. C.; Ebdon, A. M.; Christopher, O. Recent Advances in the Management of Cystic Fibrosis. *Arch. Dis. Child.* **2014,** *99*, 1033–1036.
Dodge, J. A. A Millennial View of Cystic Fibrosis. *Dev. Period Med.* **2015,** *19*(1) 9–13.
Groth, K. A.; Skakkebaek, A.; Host, C.; Gravholt, H.; Bojesen, A.; Klinefelter Syndrome—a clinical Update. *J. Clin. Endocrinol. Metab.* **2013,** *98*(1), 20–30.
Grossfeld, P.; Favier, R.; Akshmoomoff, N.; Mattson, S. Jacobsen Syndrome: Advances in our Knowledge of Phenotype and Genotype. *Am. J. Med. Genet.* **2015,** *169*(3), 239–250.
Haldeman-Englert, C. R.; Saitta, S. C.; Zackai, E. H. Chromosomal Disorders. In *Avery's Diseases of the Newborn,* 10th ed.; 2018; pp 211–223e2.
Hernandez, L. M.; Blazer, D. G. Genes, Behavior and the Social Environment. In *Moving Beyond the Nature/Nurture Debate*; The National Academy Press: Washington D.C., 2006; p 44.
Kathiresan, S.; Srivastava, D. Genetics of Human Cardiovascular Disease. *Cell* **2012,** *148*(6), 1242–1257.
Levitsky, L. L.; Luria A. H.; Hayes, F. J.; Lin, A. E. Turner Syndrome: Update on Biology and Management Across the Lifespan. *Curr. Opin. Endocrinol. Diabetes Obes.* **2015,** *25*, 65–72.
Lu, Q.; Aguilar, B. J.; Li, M.; Jiang, Y.; Chen, Y. H. Genetic Alterations of δ-Catenin/NPRAP/Neurojungin (CTNND2): Functional Implications in Complex Human Diseases. *Hum. Genet.* **2016,** *135*(10), 1107–1116.
Martin, A. M.; Weber, B. L. Genetic and Hormonal Risk Factors in Breast Cancer. *J. Natl. Cancer. Inst.* **2000,** *92*(14).
Poulter, N. Coronary Heart Disease is a Multifactorial Disease. *Am. J. Hypertens.* **1999,** 92S–95S.
Robinson, P. N.; Godfrey, M. The Molecular Genetics of Marfan Syndrome and Related Microfibrillopathies. *J. Med. Genet.* **2000,** *37*, 9–25.
Satge, D.; Nishi, M.; Sirvent, N.; Vekemans, M. A Tumor Profile in Edwards Syndrome (Trisomy 18). *Am. J. Med. Genet.* **2016,** *172C*, 296–306.
Serjeant, G. R. The Natural History of Sickle Cell Disease. *Cold Spring Harb. Perspect. Med.* **2013,** *3*, a011783.
Tartaglia, N.; Lee, C. H.; Cordeiro, L. Cognitive and Medical Features of Chromosomal Aneuploidy. *Handb. Clin. Neurol.* **2013,** *111*(1).
Tsuji, D. Molecular Pathogenesis and the Therapeutic Approach of GM2 Gangliosidosis. *J. Pharma. Soc. Japan.* **2013,** *133*(2), 269–274.
Walley, A. J.; Blakemore, A. I. F.; Froguel, P. Genetics of Obesity and the Prediction of Risk for Health. *Hum. Mol. Genet.* **2006,** 124–130.
Weinhold, B. Epigenetics: The Science of Change, Environmental Health Perspectives. *Focus* **2016,** *114*(3).
Xia, Q.; Grant, S. F. A. The Genetics of Human Obesity. *Ann. N.Y. Acad. Sci.* **2013,** *1281*, 178–190.

Whijmenga, C.; Tilburg, J.; Haeften, T. W.; Pearson, P. Defining the Genetic Contribution of Type 2 Diabetes Mellitus. *J. Med. Genet.* **2001,** *38*, 569–578.

Wright, A. D.; Wang, H.; Gurrath, M.; Koenig, G. M.; Kocak, G.; Neumann, G.; Loria, P.; Foley, M.; Tilley, L. Inhibition of Heme Detoxification Processes Underlies the Antimalarial Activity of Terpene Isonitrile Compounds from Marine Sponges. *J. Med. Chem.* **2001**, *44*, 873–885.

Wright, C. W. Traditional Antimalarials and the Development of Novel Antimalarial Drugs. *J. Ethnopharmacol.* **2005**, *100*(1–2), 67–71.

Zheng, Z.; Diamond M. I. Hungtinton Disease and the Huntingtin Protein. *Prog. Mol. Biol. Transl. Sci.* **2012,** *107*, 189–214.

CHAPTER 5

Current Perspectives and Trends in Gene Therapy and Their Clinical Trials

PRAMOD KUMAR MAURYA, NEHA SHREE MAURYA, and ASHUTOSH MANI*

Department of Biotechnology, Motilal Nehru National Institute of Technology Allahabad 211004, India

**Corresponding author. E-mail: amani@mnnit.ac.in*

ABSTRACT

The need of target-specific modifications in human genome for the purpose of treating the genetic diseases has always been there. Gene therapy is known as a method of altering and mutating the genes to be used for therapeutic purposes. Approximately, 2600 gene therapy clinical trials have been either accomplished, while some are currently ongoing or have been sanctioned globally, till date. This therapy is nowadays very popular because of the advances in the genetic engineering methods and bioengineering technology which has improved the efficiency of manipulation through various vectors both in vivo and in vitro.

For many years, viral vectors particularly retroviruses and adenoviruses have been majorly used as vehicle to transfer genes into human cells. The current gene therapy directed clinical trials are focused on advancements in genome editing technologies like CRISPR-cas9 and cancer treatment with engineered antigen receptor T cells. The major countries which are participating in gene therapy- based clinical trials are the United States, China, Europe, and Australia.

This approach is broad in experimental sense and still needs various new developments and is still in its experimentally driven phase for various genetic disorders. This chapter aims on focusing the trends which are being followed up by the researchers nowadays for current gene therapy-based clinical trials.

5.1 INTRODUCTION

5.1.1 GENE THERAPY AND ITS TRIALS

The concept of gene therapy trials began in 1989 by Rosenberg et al. when retroviruses were used for gene transduction in human tumor infiltrating lymphocytes where they reinfused it in five patients with advanced melanoma (Adams et al., 2018). The study provided the first landmark evidence that patients could be treated with genetically modified human cells without any harm. This study formed the basis for treatment of serious inherited diseases that were incurable earlier. After this landmark study, first clinical trial was performed in 1990 with two girls affected with adenosine deaminase (ADA)—severe combined immunodeficiency (SCID) using T cell in which ADA gene was transferred by retrovirus. The number of clinical trials increased during 1990s (Androulla et al., 2018; Blaese et al., 1995; Boudes, 2014).

On September 17, 1999, a gene therapy treatment caused death of an 18 year old Jesse Gelsinger who had ornithine transcarbamylase deficiency. In January 2000, Food and Drug Administration halted several clinical trials till February 2005 and placed several research restrictions to conduct gene therapy trials (Bowles et al., 2012; June et al., 2016). In 2000, the researchers from France also successfully treated children affected from SCID-X1 which was due to early block in lymphocyte differentiation (Cavazzana-Calvo et al., 2000). In April 2006, a Swiss-German gene therapy clinical trial was successfully conducted in treating granulomatous disease which affects phagocytes (Cockburn, 2004). In September 2006, a gene therapy clinical trial involving genetically engineered lymphocytes was reported successfully to treat two patients having metastatic melanoma (Couzin & Kaiser, 2005). In October 2006, Parkinson's disease was successfully treated in 12 patients by using an adeno-associated virus vector having neurturin gene which was introduced into each patient (Fire et al., 1998). In 2010, the recombinant adeno-associated viral vectors (rAAV) expressing a retinal pigment protein RPE65 were used to treat leber congenital amaurosis (LCA), a form of congenital blindness in which the vision was reported maintained for up to 2 years (Ginn et al., 2012). In 2011, the researchers in Philadelphia successfully treated three patients having chronic lymphocytic leukemia (CLL) by using an autologous T cells having chimeric antigen receptors (CARs) for B cell antigen CD19 (Grigor et al., 2017).

5.2 EMERGING TRENDS IN GENE THERAPY TRIALS

5.2.1 CAR T CELLS

Immune system of the body is known to provide immunity against various types of infections and diseases by recruiting various cells which are involved in defense systems. Lymphocytes are one of those kinds of cells which play important role in providing immunity. Lymphocytes have been categorized as follows:

- T lymphocytes which are differentiated as T_h and T_c cells. T_h have role in activating B cells and T_c cells combat infection and directly kill cells which are infected inside the body.
- B lymphocytes produce antibodies to fight against infection.
- Natural killer cells attack and remove the virus infected cells from the body.

Immunotherapy is the type of treatment in which the cells of the immune system are involved in fighting against various diseases. This therapy is especially used to fight against various forms of cancers. Adoptive T Cell Transfer (ACT) is known for infusion of lymphocytes method to produce antiviral, anti-inflammatory, and antitumor state. Chimeric antigen receptor T (CAR T) cell-based immunotherapy is one of the forms of ACT developed for therapy against cancer.

CAR T cells are genetically engineered cells which possess receptors CD3ζ (main molecular marker), CD28, 4-1 BB as the co molecular markers on the T cells. It depends on the safe, efficient, and stable gene transfer platforms. Tisagenlecleucel is approved for the treatment of patients with B-ALL (during refractory or in second or later relapse). Axicabtagene ciloleucel is approved for the treatment in adults with relapsed large B cell lymphoma after more than 2 lines of systemic therapy Hacein-Bey-Abina et al., 2003).

Properties of the CAR T cells are as follows:

- Avidity is controllable
- MHC-independent recognition of tumor targets

- Targets proteins and glycans associated molecules typically expressed on tumor cell's surface
- Decade long persistence of the therapy
- Effective killer of tumor cells

The domain which is present extracellularly in the CAR consists of the antigen binding moiety that could be:

a) Single-chain fragment variable (scFv) which is derived from antibodies;
b) Selected human Fab fragment from phage display libraries; or
c) Cognate receptor that are engaged through their natural ligands.

The antigen binding domain is connected with the transmembrane domain with the help of the spacer. Transmembrane domain is connected with CD28 receptor which is present intracellularly. First, signal for T cell activation and function is provided through CD3ζ, component of intracellular domain. The increased secretion of cytokines (IL-2) is marked by the co-stimulatory signaling between CD28 or 4- 1BB and the in vivo persistence and expansion of T cells. CAR T cells which are optimized design of CARs are known to induce the universal cytokine production for tumor cell killing by producing transgenic product such as IFN-γ or IL-12. Steps which are involved in the CAR T cell-based therapy are as follows (Fig. 5.1):

a. **Collection of T cell from patients** through apheresis in which WBCs and plasma are isolated from the blood of the patient and the remaining blood is returned to the patient's body.
b. **T cells reprogramming** in the laboratory for genetically engineering cells through various viral and nonviral transfer methods which will express CARs on their surface.
c. **Increasing genetically engineered T cells** by culturing them in laboratory where they will get expanded to a considerable amount and will be sent back to the hospital where patient will be treated.
d. **Infused CAR T cells** in the body will attract the cancer cells and will destroy them. They will remain in the blood for some months to prevent cancerous cells from returning (Hirai et al., 1997).

Benefits associated with CAR T cells-based therapy are as follows:

- Helps in destroying the tumor cells.
- CAR T cells keeps on growing and dividing inside the body.

- Long-lasting effect over many months or years.
- Works best when conventional treatment gets failed.

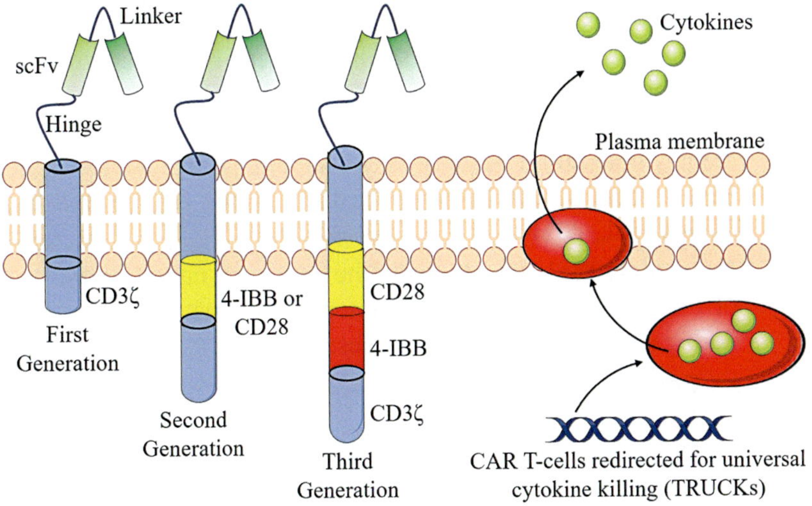

FIGURE 5.1 Diagrammatic representation of structure for CAR (Androulla et al., 2018).

The treatments which are involved for cancer treatment such as chemotherapy and radiation-based therapy kills normal cells also along with the cancerous cells. Although, these side effects are not correlated with the CAR T cells-based therapy, while some of the side effects coupled with the CAR T cells-based therapy are:

a. **Cytokine Release Syndrome:** The sign of working treatment correlated with CAR T cells therapy is also one of its main side effects. As the cells grow and divide inside the body they produce chemical messengers (cytokines) that helps in launching the immune system against the cancer cells. Cytokines are necessary for killing of tumor cells but an increased amount inside the body then what is required can cause flu like symptoms, such as high fever, body ache, chills, nausea, fatigue. These symptoms will occur within few days of treatment and will go away after 1 to 2 weeks. Less often CRS may have side effects more severely in some patients, such as:

   I. Low blood pressure
   II. Irregular heart rhythm

III. Low oxygen in the blood
IV. Heart failure
V. Renal insufficiency associated with poor kidney function
VI. Capillary leakage in which proteins and fluid leak out.

Medicines like steroids and tocilizumab can help to control these symptoms, but a small number of patients need to be treated in the hospital. Most of the symptoms are reversible with CAR T cell therapy but still the associated risk should not be underestimated. Several deaths have been reported during the CAR T cell trials (Kalos et al., 2011).

b. **Neurologic Toxicities:** Includes the frequency, nature, and severity of the neurologic effects. Common symptoms which are found during neurologic toxicity includes:
   - Confusion
   - Language impairment
   - Hallucinations
   - Involuntary muscle twitching
   - Delirium
   - Seizures

   Neurotoxicity has shown to be changeable in many cases and the symptoms have vanished without any long-term effects.
c. **B cell Aplasia:** It is a condition in which B cell are either absent or are present in very low numbers because of the successful CD-19-specific CAR T cell treatment. Therefore, B cell aplasia serves as an important indicator for this therapy. Normal and cancerous both cells are destroyed in response to this therapy which targets the antigens present on the surface of B cells.
d. **Macrophage activation syndrome (MAS):** Symptoms which are found with macrophage activation includes high levels of c-reactive protein, ferritin, and d-dimer. Tocilizumab drug has been effectively used for treating MAS by blocking IL-6 receptor (Kalos et al., 2011).

5.2.2 CLINICAL TRIAL RESULTS FOR CAR T CELL THERAPY

In recent years, CAR T cells have been successfully used to aim cell-surface antigen in cancer cells but now this strategy has been updated to improve the cytotoxic T cells persistence and potential- specific monoclonal antibodies against tumor cells. The Kymriah was first FDA approved CAR T cell product which was developed against acute lymphoblastic leukemia in 2017. The complete response rate for the refractory and relapsed B cell acute lymphoblastic leukemia (r/r B-ALL) ranges from 70% to 90% CAR T cell-based therapy. In some cases relapse had occurred when tumor cell loose expression of CD-19 marker which are sometimes known as "escape variants of CD-19." Till now September, 2018, total numbers of clinical trials for CAR T cell therapy worldwide are 316. Countries of East Asia, North America, and the United States are the leading members in the studies based on CAR T cells-based therapy (data were accessed using clinicaltrials.gov website). The data representing the status of clinical trials worldwide for CAR T cells therapy are represented through the following data:

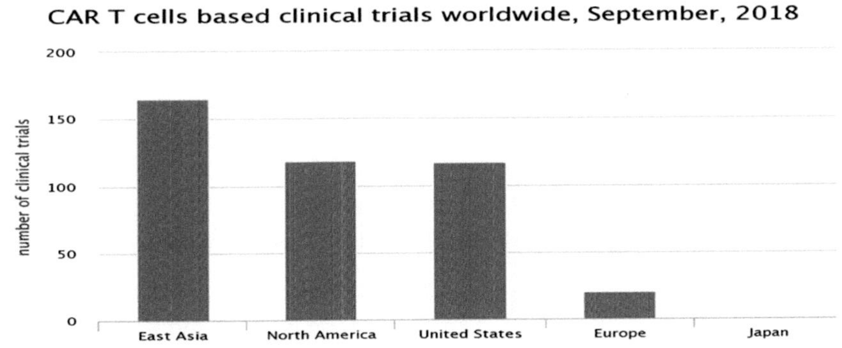

FIGURE 5.2 Geographic localization of all clinical trials worldwide for CAR T cell therapy [39].

5.2.3 CRISPR/CAS9 SYSTEM-BASED GENOME EDITING

In recent time, the field of human genetic engineering has been revolutionized by the development of the new genome editing tool known as CRISPR Cas9 (clustered regularly interspaced short palindromic repeats/CRISPR-associated protein-9 nuclease). This system in bacteria and archaea naturally acts as an adaptive defense mechanism against bacteriophages and plasmids which contains foreign nucleic acids with them. For the purpose of editing DNA, programmable DNA-binding nucleases such as and TALENs (transcription-activator like nucleases) and ZNFs (zinc finger nucleases) were used before the establishment of the CRISPR/Cas9 which is RNA-guided engineered nuclease. Till now, 3 CRIPSR/Cas systems have been identified so far out of which type II which involves Cas9 proteins has been mostly studied.

Immunity through CRISPR/Cas acts in the following three steps:

I. **Adaptation:** Phase involves incorporation of short foreign sequences (protospacers) of bacterial or archaeal origin into the genome of host as repeat-spacer motifs.
II. **Expression with Maturation:** Precursor CRISPR RNAs (pre-crRNAs) are generated by the transcription of sequences which contains the repeat-spacer motifs, after maturation pre-crRNAs becomes crRNA molecules.
III. **Interference**: Recognition and targeting of the invading genetic material (Tiemann & Rossi, 2009).

Type II CRISPR/Cas9 system is capable of cleaving dsDNA and has been originated from *Streptococcus thermophilus*. The disruption of DNA at specific sites requires crRNA and **protospacer adjacent motif** (PAM) which is a DNA sequence of short length. Fusion of crRNA to the (tracrRNA) **trans-activating CRISPR RNA** generates **single guide RNA** (sgRNA). sgRNA guides the recruitment of Cas9 nuclease to the specific genomic locations through Watson-Crick's standard base pairing principle.

CRISPR/Cas9 complex drives the gene editing in two different ways:

a. **Non homologous end joining** (NHEJ) mechanism helps in repairing dsDNA breaks which is an error prone process and it causes small **InDel** (insertion and deletion).
b. Repairing of dsDNA breaks by **homology-directed repair** (HDR) which introduces desirable sequence changes (Keyhani et al., 2010).

CRISPR/Cas9 zygote editing: Purpose behind this method of editing is to introduce the changes in each and every cells of the genome of an organism including germ line. This approach can provide lasting solution of genetic diseases as the changes in the genome will be transferred to the subsequent generations.

CRISPR/Cas9 component involves:

a. HDR template;
b. sgRNA; and
c. Cas9 messenger RNA or protein.

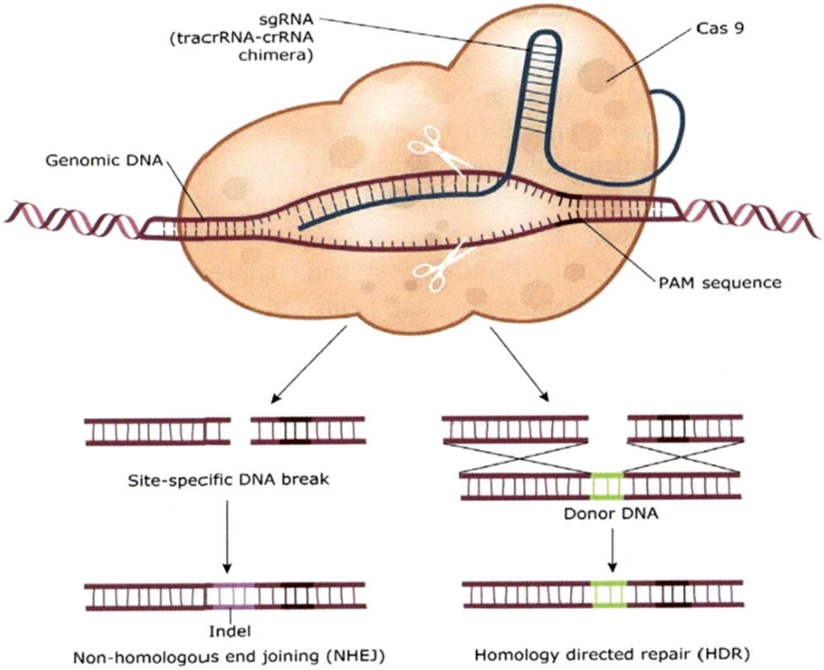

FIGURE 5.3 CRISPR/Cas9-mediated genome editing.

Source: Adapted from Ref. [35].

Wu et al. were one of the first who used genome editing through this method in mouse embryo to cure a disease-causing mutation. Mutation in the CRYGC gene was focused which was dominant and was associated with loss-of-function. This mutation causes cataract which is associated with decreased vision and clouding of the eye lenses. Inside the zygote dominant CRYGC allele was targeted by injecting the Cas9 mRNA and sgRNA, which leads to the mutation correction with the wild type allele and where used template was homologous chromosome.

This method of zygote editing was used for human embryos also which has generated controversy among scientists and public because of the ethical issues. Liang et al. had reported in their study where they had modified beta-hemoglobin gene using human's tripronuclear zygote which is responsible for β-thalassemia. However, some unwanted modifications were also observed which has raised many questions.

Limitations of the CRISPR/Cas9 System

a. For recognizing and cleaving the target sequence it is not required that guide RNA should completely match the target sequence. This characteristic feature of the system sometimes shows off-target effects which is of major concern. It is associated with off-target recognition and unpredictable mutations inside the organism.
b. Another reason of concern is delivery of Cas9 and sgRNA. For introduction of CRISPR/Cas9 system components different viral and nonviral vectors are used. Lentiviral and adenoviral vectors are very common and effective tools. But several drawbacks are associated viral vectors such as cargo capacity is limited and confinement of tissue tropism to specific organ (Kruminis-Kaszkiel et al., 2018).

5.2.4 RNA INTERFERENCE

RNAi has already been proved successful for gene silencing and also for analysis of gene functions. RNAi is a novel emerging therapy for gene suppression in which si-RNA targets the sequence- specific suppression of target gene expression. Among all other gene suppression therapies, it has advantage of si-RNA stability and adequate delivery of si-RNA (Kubowicz et al., 2013).

In 1997, Andrew Fire and his colleagues first time observed the RNA interference phenomenon by studying the introduction of dsRNA into C. elegans but the existence of RNAi to down regulate a gene expression came in 2006 when the noble prize in physiology and medicine was received by Fire and Mello (Laitala-Leinonen, 2010; McGovern, 2012).

In 2012, Santel et al. have successfully silenced the gene expression of the protein kinase N3 which results in efficient inhibition of tumor growth and metastasis in lung carcinoma. Till now, many successful experimental studies and clinical trials conducted on si-RNA approach have proven its potential in specific and systemic cancer gene therapies (Montgomery, 2006).

In pharmaceutical industry, RNAi has most importance due to its unique ability of self-delivering without any additional delivery vehicle. Currently, Pharma companies are observing the clinical trial status of si-RNA studies in order to develop RNAi therapeutics (Morgan et al., 2006). From pharmaceutical point of view, the clinical research on si-RNA studies has been majorly focused on macular degeneration, cancer, and antiviral therapy but also observed in studies related to asthma, hypercholesterolemia, inherited skin disorders, and amyloidosis. To date, approximately 20 clinical trials have been conducted on RNAi-based therapeutics (http://www.ukctg.nihr.ac.uk). The first RNAi therapeutics was developed in the 2000s which was successful. After phase 3 clinical trial, Alnylam Pharmaceuticals has announced patisiran as an RNA interference drug to treat patients suffering from a life-threatening genetic disorder named hereditary transthyretin amyloidosis. Patisiran has become the first RNAi therapeutic on the US market. Patisiran is a novel type of medicine that uses RNAi to suppress a gene from producing disease-causing proteins by degrading target mRNA with use of dsRNA (Nayak & Herzog, 2010).

5.3 CURRENT STATUS OF CLINICAL TRIAL ASPECTS IN GENE THERAPY

During 2000s, beside the United States, many other regulating agencies in different countries had maintained their own clinical trial databases on gene therapy which were either freely available or available on request. The availability of these national, regional, and global databases along with clinical trial regulations bring transparency and increased the information about gene therapy clinical trials (Nayerossadat et al., 2012).

The variation in policy on freely available data by regulatory agency or authorities could be seen among different countries. The National Laboratory of Medicine at NIH maintains a web-based database for all completed and ongoing clinical trials in the United States (Ott etal., 2006). The National Institute for Health Research in the United Kingdom maintained a UK clinical trial database to provide the information on clinical trials (Pang et al., 2018). The Federal Ministry of Education and Research in Germany also maintained a German clinical trial database which is a freely accessible database managed by nonprofit organization (Raper et al., 2003).

5.3.1 NUMBER OF TRIALS APPROVED/INITIATED PER YEAR

After first gene therapy trial by Rosenberg et al.,[1] the number of clinical trial increased with the development of gene therapies. The number of trials did not increase steadily from 1990 to 2017. In year 2003, 2007, 2009, 2011, there was drop in trials due to some adverse reports on gene therapy events. However, an increase in gene therapy clinical trials was observed from 2012 to 2015 and we also noticed an increase in clinical trials from 2016 to 2018. The year 2016 was the most promising year for gene therapy but year 2017 also witnessed big gene therapy developments. Indeed in 2017, the two pioneer gene therapy treatments came into USA market after approval of FDA. These are Kymriah and Yescarta which uses a patient's own T cell to fight against rare types of cancer (Riviere & Sadelain, 2017; Rosenberg et al., 1990).

5.3.2 COUNTRIES WHERE GENE THERAPY CLINICAL TRIALS WERE CONDUCTED

Gene therapy clinical trials have been performed in 38 countries from all 5 continents. Since 2012, seven new countries have started trials which are Argentina, Burkina Faso, Gambia, Kenya, Kuwait, Senegal and Uganda. The United States undertook 64.9% of all gene therapy clinical trials which was 65.1% in 2012 and Europe undertook 23.2% of total trials which was 28.3% in 2012 and Asia undertook 6.5% with slight growth which was 3.4% in 2012. The United States continues to maintain top rank for conducting globally 1643 trials due to huge investment, high funding organization like NIH. After the United States, the United Kingdom also offers friendly environment for entrepreneurs and tax relief. In Asia, Japan and China also increased growth in gene therapy clinical trials (Nayerossadat et al., 2012; Riviere & Sadelain, 2017; Rosenberg et al., 1990; http://www.rush.edu/webapps/MEDREL/servlet/NewsRelease?id=802; Samantha et al., 2018).

5.3.3 DISEASES TARGETED BY GENE THERAPIES

Cancer was the most common disease for which majority of clinical trials (65.0%) have been conducted. Total 1688 trials have been conducted to treat gynecological, lung, skin, genitourinary, sarcoma, head and neck, gastrointestinal, and neurological tumors. Various therapies include transferring tumor suppressor genes, immune therapy, CAR T cell therapy, oncolytic virotherapy, and gene-mediated enzyme prodrug therapy. The CAR T cell therapy has shown impressive results for lymphoid abd myeloid leukemias (Nayerossadat et al., 2012; Santel et al., 2010; Savic & Schwank, 2015).

The second most popular disease types for gene therapy were monogenic diseases (11.1%) in which functional gene is transferred into dividing stem cells to treat the disease. Total 287 trials have been identified for inherited monogenic diseases in which 16.3% of trials have been targeted for immune-deficiency disorders and 12.5% of trials were for targeting cystic fibrosis. Another most common disease is cardiovascular disease for which 6.9% of clinical trials were conducted under cardiovascular gene therapy category (Nayerossadat et al., 2012; Simonelli et al., 2010; hurtle-Schmidt & Lo 2018).

5.3.4 CLINICAL TRIALS PHASE AND STATUS FOR GENE THERAPIES

Majority of gene therapy-based clinical trials were in early phase of development as more than three quarter of gene therapy clinical trials performed are phase I or phase I/II till now. The phase I and phase I/II represent 77.7% of all gene therapy clinical trials, while phase II clinical trials make up 17.1% of trials and phase II/III and phase III represent only 4.8% of the total trials. Most of the trials in early phase have to cope with regulatory approval of therapy for clinical practice beside several obstacles such as funding, health issues, and patient's access (Nayerossadat et al., 2012; Raper et al., 2003).

5.3.5 VECTORS USED FOR GENE THERAPY

Different vectors and delivery systems have been used in gene therapy clinical trials. Viral vectors are more popular than nonviral vectors currently; however, use of nonviral gene therapy is increasing consistently. Adenovirus and retrovirus are most commonly used in gene therapy among all successfully used viral vectors. The use of retroviral vectors currently has declined to 17.4% of total clinical trials which was 19.7% in 2012, while lentiviral vectors exhibit an increased used in gene therapy having 7.3% share of total clinical trials which was 2.9% in 2012. Adeno-associated viruses also followed same trend as it increased to 7.6% of total trials which was 4.9% in 2012. Adenoviruses are most popular vectors, having 20.5% share of all trials which was 23.3% in 2012. They can carry large DNA inserts up to 35 kb but for clinical applications, its insert capacity is very low (its capacity is very low: what does it mean?).

Nonviral vector delivery systems have been developed due to safety issues and also for small size of therapeutic DNA. Nonviral vectors are less efficient than viral vectors but they are better in terms of availability and cost-effectiveness. Viral vectors may cause undesirable change by stimulating host immune response. Naked DNA has been used in 16.5% of clinical trials which was 18.3% in 2012. The most commonly used methods for delivering nonviral vector systems are liposomes fusion, particle bombardment, electroporation, and ultrasound. Redesigned AAVs makes the next generation of adeno-associated viruses (Nayerossadat et al., 2012; https://www.drks.de/drks_web; http://www.clinicaltrials.gov; Yla-Herttuala & Baker, 2017; Thomas et al., 2003).

5.4 CONCLUSION

Advancements in gene therapy techniques along with increased investments in gene therapy clinical trials has paved way for adding new aspects in therapeutic manufacturing and commercialization industry. Since the first experiments conducted in the year 1998 about two decades later, in 2017 FDA approved two gene therapy products namely Kymriah and Yescarta as pioneer treatment against bone marrow cancer and lymphoma by using CAR T cell gene therapy-based clinical trials.

Besides CAR T cells, RNAi, CRISPR Cas9, and engineered AAVs, there are other molecular tools which are currently in bench stage of clinical trials. Viral vectors have viral toxicity but mi-RNA and si-RNA toxicity has not been studied in mammalian cells till now. CRISPR Cas9 can repair a gene in situ and could be a permanent germ-line treatment tool against a genetic disease. Main goal of current gene therapy is to control gene expression and hopefully more efficient methods will be evolved for making it more safe, efficient, and economic.

KEYWORDS

- **cancer**
- **prodrug therapy**
- **leukemia**
- **gene therapy**
- **clinical trials**

REFERENCES

Adams, D.; Gonzalez-Duarte, A.; O'Riordan, W. D.; et al. Patisiran, an RNAi Therapeutic, for Hereditary Transthyretin Amyloidosis. *N. Engl. J. Med.* **2018**, 379(1), 11–21.

Androulla, M. N.; Lefkothea, P. C. CAR T-cell Therapy: A New Era in Cancer Immunotherapy. *Curr. Pharm. Biotechnol.* **2018**, 19(1), 5–18.

Blaese, R. M.; Culver, K. W.; Miller, A. D.; et al T Lymphocyte-Directed Gene Therapy for ADA-SCID: Initial Trial Results After 4 Years. Science 1995, 270, 475–480.

Boudes, P. Gene Therapy as a New Treatment Option for Inherited Monogenic Diseases. *Eur. J. Int. Med.* **2014**, *25*, 31–36.

Bowles, D. E.; McPhee, S. W.; Li, C.; et al. Phase 1 Gene Therapy for Duchenne Muscular Dystrophy using a Translational Optimized AAV Vector. *Mol. Ther.* **2012**, *20*(2), 443–455.

Cavazzana-Calvo, M.; Hacein-Bey, S.; de Saint Basile, G.; et al. Gene Therapy of Human Severe Combined Immunodeficiency (SCID)- X1 Disease. Science 2000, 288, 669–672.

Cockburn, I. M. The Changing Structure of the Pharmaceutical Industry. *Health Aff.* **2004**, *23*(1), 10–22.

Couzin, J.; Kaiser, J. As Gelsinger Case Ends, Gene Therapy Suffers Another Blow. *Science* **2005**, 307, 1028.

Fire, A.; Xu, S.; Montgomery, M. K.; Kostas, S. A.; Driver, S. E.; Mello, C. C. Potent and Specific Genetic Interference by Double-Stranded RNA in Caenorhabditis Elegans. *Nature* **1998**, 391, 806–811.

Ginn, S. L.; Alexander, I. E.; Edelstein, M. L.; et al. Gene Therapy Clinical Trials Worldwide to 2012 – an Update. *J. Gene. Med.* **2013**, 15, 65–77.

Grigor, E. J. M.; Fergusson, D. A.; Haggar, F.; et al. Efficacy and Safety of Chimeric Antigen Receptor T-Cell (CAR-T) Therapy in Patients with Haematological and Solid Malignancies: Protocol for a Systematic Review and Meta-Analysis. *BMJ Open* **2017**, 7, e019321.

Hacein-Bey-Abina, S.; von Kalle, C.; Schmidt, M.; et al. A Serious Adverse Event After Successful Gene Therapy for X-linked Severe Combined Immunodeficiency. *N. Engl. J. Med.* **2003**, 348, 255–256.

Hirai, H.; Satoh, E.; Osawa, M.; et al. Use of EBV-Based Vector/HVJ-Liposome Complex Vector for Targeted Gene Therapy of EBV-Associated Neoplasms. *Biochem. Biophys. Res. Commun.* **1997**, 241, 112–118.

June, C. H.; O'Connor, R. S.; Kawalekar, O. U.; Ghassemi, S.; Milone, M. C. CAR T Cell Immunotherapy for Human Cancer. Science 2016, 359(6382), 1361–1365.

Kalos, M.; Levine, B. L.; Porter, D. L.; et al. T Cells with Chimeric Antigen Receptors have Potent Antitumor Effects and Can Establish Memory in Patients with Advanced Leukemia. *Sci. Transl. Med.* **2011**, 3, 95ra73.

Keyhani, S.; Wang, S.; Hebert, P.; et al. US Pharmaceutical Innovation in an International Context. *Am. J. Public Health* **2010**, *100*(6), 1075–1080.

Kruminis-Kaszkiel, E.; Juranek, J.; Maksymowicz, W.; Wojtkiewicz, J. CRISPR/Cas9 Technology as an Emerging Tool for Targeting Amyotrophic Lateral Sclerosis (ALS). *Int. J. Mol. Sci.* **2018**, 19, 906.

Kubowicz, P.; Żelaszczyk, D.; Pękala, E. RNAi in Clinical Studies. *Curr. Med. Chem.* **2013**, *20*(14), 1801–1816.

Laitala-Leinonen, T. Update on the Development of microRNA and siRNA Molecules as Regulators of Cell Physiology. Recent Pat. *DNA Gene Seq.* **2010**, 4, 113–121.

McGovern, V. Getting Grants. *Virulence* **2012**, *3*(1), 1–11.

Montgomery, M. K. RNA Interference: Unraveling a Mystery. *Nat. Struct. Mol. Biol.* **2006**, 13, 1039–1041.

Morgan, R. A.; Dudley, M. E.; Wunderlich, J. R.; et al. Cancer Regression in Patients After Transfer of Genetically Engineered Lymphocytes. *Science* **2006**, 314, 126–129.

National

Nayak, S.; Herzog, R. W. Progress and Prospects: Immune Responses to Viral Vectors. *Gene Ther*. **2010**, *17*(3), 295–304.

Nayerossadat, N.; Maedeh, T.; Ali, P. A. Viral and Non-Viral Delivery Systems for Gene Delivery. *Adv. Biomed. Res.* **2012**, 1, 27.

Ott, M. G.; Schmidt, M.; Schwarzwaelder, K.; et al. Correction of X-linked Chronic Granulomatous Disease by Gene Therapy, Augmented by Insertional Activation of MDS1- EVI1, PRDM16 or SETBP1. *Nat. Med.* **2006**, 12, 401–409.

Pang, Y.; Hou, X.; Yang, C.; et al. Advances on Chimeric Antigen Receptor-Modified T-Cell Therapy for Oncotherapy. *Mol. Cancer* **2018**, 17, 91.

Raper, S.E.; Chirmule, N.; Lee, F. S.; et al. Fatal Systemic Inflammatory Response Syndrome in a Ornithine Transcarbamylase Deficient Patient Following Adenoviral Gene Transfer. *Mol. Genet. Metab.* **2003**, 80, 148–158.

Riviere, I.; Sadelain, M. Chimeric Antigen Receptors: A Cell and Gene Therapy Perspective. *Mol. Ther.* **2017**, 25, 1117–1124.

Rosenberg, S.A.; Aebersold, P.; Cornetta, K.; et al. Gene Transfer into Humans – Immunotherapy of Patients with Advanced Melanoma, using Tumor-Infiltrating Lymphocytes Modified by Retroviral Gene Transduction. *N. Engl. J. Med.* **1990**, 323, 570–578.

Rush University Medical Center and UCSF Researchers Found Gene Therapy Appeared to Reduce Symptoms of Parkinson's Disease by 40 Percent. Press Release from Rush University Medical Center. http://www.rush.edu/webapps/MEDREL/servlet/NewsRelease?id=802 (accessed Oct 16, 2006).

Samantha, L. G.; Anais, K. A.; Ian, E. A.; Michael, E.; Mohammad, R. A. Gene Therapy Clinical Trials Worldwide to 2017: *An Update. J. Gene. Med.* **2018**, 20, e3015.

Santel, A.; Aleku, M.; Röder, N.; et al. Atu027 Prevents Pulmonary Metastasis in Experimental and Spontaneous Mouse Metastasis Models. *Clin. Cancer Res.* **2010**, 16, 5469–5480.

Savic, N.; Schwank, G. Advances in Therapeutic CRISPR/Cas9 Genome Editing. *Transl. Res.* **2015**, 168, 15–21.

Simonelli, F.; Maguire, A. M.; Testa, F.; et al. Gene Therapy for Leber's Congenital Amaurosis is Safe and Effective Through 1.5 Years After Vector Administration. *Mol. Ther.* **2010**, 18, 643–650.

Thomas, C., Ehrdart, A. & Kay, M. Progress and problems with the use of viral vectors for gene therapy. *Nat Rev Genet.* **2003**, 4, 346–358.

Thurtle-Schmidt, D. M.; Lo, T. Molecular Biology at the Cutting Edge: A Review on CRISPR/CAS9 Gene Editing for Undergraduates. Biochem. *Mol. Biol. Educ.* **2018**, 46, 195– 205.

Tiemann, K.; Rossi, J. J. RNAi-Based Therapeutics–Current Status, Challenges and Prospects. *EMBO Mol. Med.* **2009**, *1*(3), 142–151.

University Medical Centre Freiburg and the German Cochrane Center, German Clinical Trials Register. https://www.drks.de/drks_web.

US National Library of Medicine, Clinical Trials Website. http://www.clinicaltrials.gov.

Yla-Herttuala, S.; Baker, A. H. Cardiovascular Gene Therapy: Past, Present, and Future. *Mol. Ther.* **2017**, 25, 1095–1106.

CHAPTER 6

Immunogene Therapy in Cancer

SREERANJINI PULAKKAT and VANDANA PATRAVALE*

Department of Pharmaceutical Sciences and Technology, Institute of Chemical Technology, Matunga (E), Mumbai 400019, Maharashtra, India

**Corresponding author. E-mail: vb.patravale@ictmumbai.edu.in; vbp_muict@yahoo.co.in*

ABSTRACT

Immunotherapy has long been investigated as a potent, alternative approach to conventional cancer therapy but the clinical translations are limited. Immuno-oncology and genomics have experienced vast advancements in the recent past that this proliferation of knowledge gave birth to the field of cancer immunogene therapy. The pioneering works on immune checkpoint pathways and their inhibition to allow T-lymphocytes to effectively eradicate cancer cells have revolutionized the field of immunotherapy and was duly acknowledged by conferring the 2018 Nobel Prize in Physiology or Medicine to James P. Allison and Tasuku Honjo. Further, the genetic manipulation of immune cells, either in situ or ex vivo, has opened new avenues to induce antitumor immune responses. The different approaches explored in immunogene therapy including ex vivo manipulation of T-lymphocytes, transferring immunostimulatory genes to tumor cells and antigen-presenting cells (APCs), in vivo genetic modulation using viral and nonviral vectors, etc. have been discussed in this chapter. The underlying rationales and the main studies that were translated to clinical trials in each approach have been briefly discussed. Although a large number of promising preclinical investigations have been reported, very few clinical trials have resulted in favorable outcomes and the road to significant clinical impact is still long. A better understanding of the key molecules in tumor immunomodulation and their mechanisms along with the use of state-of-the-art genetic engineering tools can enable the rational development of successful cancer immunogene therapies.

6.1 INTRODUCTION

Cancer remains as the second leading cause of death globally, accounting for an estimated 9.6 million deaths in 2018 alone. Although the incidence of cancer is on the rise, the mortality due to cancer has declined owing to the enormous amount of research in the field of cancer therapy. Cancer is a generic term referring to a group of diseases characterized by the rapid growth of abnormal cells beyond their boundaries and in most cases, invasion to adjoining parts of the body, that is, metastasis ("Cancer" n.d.). Cancer pathogenesis involves multiple processes through which the normal cells progressively acquire essential traits called "hallmarks of cancer" via genetic mutations and epigenetic alterations. A better understanding of the complex interactions between cancer cells and the surrounding microenvironment has established the ability of transforming cells to evade immune system as an important hallmark and the role of immune inflammatory cells in tumor promotion (Hanahan and Weinberg 2011). This led to exciting developments at the interface of oncology and immunology in the past decade. There is a dynamic crosstalk between immunology and cancer pathogenesis by which the immune system can protect the host against tumor development (immunosurveillance) as well as promote tumor growth (tumor immune escape). This concept of "cancer immunoediting" involves three essential phases: elimination, equilibrium, and escape, designated as the "three E's". In the elimination or the "immunosurveillance" phase, the nascent transformed cells are eradicated initially by immune effector cells such as natural killer (NK) cells, macrophages, and dendritic cells (DC), etc. via production of interferons (IFNs) and chemokines. The activated DCs mature and migrate to the draining lymph node where $CD4^+$T-helper (Th)1 T- lymphocyte (T cell) activation and $CD8^+$ cytotoxic T cell proliferation occurs. This leads to T cell homing to the tumor site and further elimination of the cancer cells. Cancer cells generate "neoantigens" that favor tumor cell recognition and production of danger signals that activate innate and adaptive immune responses. In addition, tumor cell death occurring in the tumor microenvironment can further release tumor antigens and pro-inflammatory mediators leading to an antitumor immune response. However, some tumor cells and cancer stem cells have decreased immunogenicity and become resistant to immune effector cells in the equilibrium phase, the longest of the three phases. In this phase, cancer immunoediting occurs and eventually, the resulting resistant variants finally enter the escape phase in which they evade the immune system and proliferate (Gajewski, Schreiber, and Fu 2013; Kim, Emi, and Tanabe 2007; Croci et al. 2007; Dunn, Old, and Schreiber 2004; Muenst et al. 2016). A schematic portrayal of the immunoediting of cancer is given in Figure 6.1.

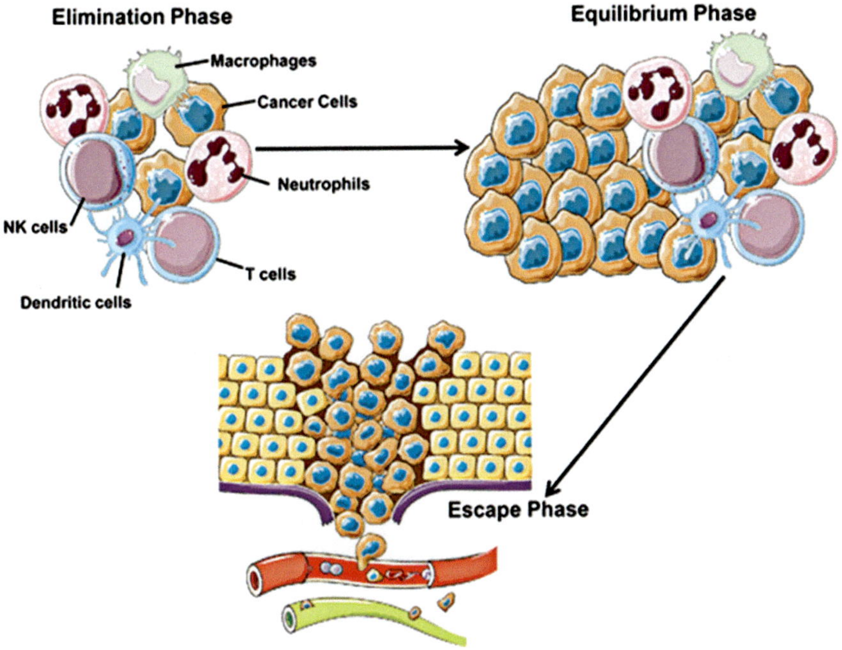

FIGURE 6.1 Schematic of immunoediting of cancer. (Reprinted with permission from Springer Nature Customer Service Centre GmbH: Springer Nature, Contribution of ER Stress to Immunogenic Cancer Cell Death by Abhishek D. Garg, Dmitri V. Krysko, Jakub Golab et al., 2012).

Several immune regulatory mechanisms adopted by the tumor cells to inhibit antitumor immune responses include lowering immunogenicity by overexpression of inhibitory receptors (called "immune check-points") and altering tumor antigen presentation, emergence of antigen loss tumor variants, recruitment of immunosuppressive cells, formation of an immunosuppressive tumor microenvironment by production of immunosuppressive mediators, increased fibrogenesis, etc. (Drake, Jaffee, and Pardoll 2006; Spranger and Gajewski 2018). Thus, the complex role of immune system in the cancer progression involves immunosuppressive functions of the multiple cell populations along with the pro-tumoral influence of the inflamed tumor tissue. Considering this, there is a need for parallel targeting of diverse cells and therapeutic manipulation of immune regulatory mechanisms to achieve an effective antitumor response.

The various strategies that have been employed to generate or strengthen immunity against tumors can be broadly classified into two main groups: passive therapies where patients are administered with a molecule (e.g., cytokines or antibodies) and active therapies which stimulate a patient's immune system to effectively respond to the tumor. In designing various active strategies, gene therapy has emerged as a promising approach enabling effective antitumor response and the driving force behind an exciting field of research called cancer immunogene therapy. Cancer immunogene therapy mainly involves the genetic manipulation of human cells and the complex interactions between tumor cells, antigen-presenting cells (APCs), T cells, NK cells, etc. to stimulate or enhance antitumor immunity. Intricate pathways involved in antigen presentation, cytokine production, lymphocyte activation and recruitment, and eventual elimination of tumor cells present several opportunities to modulate the immune response. The potential strategies that have been investigated for immunogene therapy of cancer include:

- Enhancing the immunogenicity of the tumor, for example, introducing genes that encode foreign antigens and co-stimulatory molecules like B7-1.
- Enhancing immune cells to intensify the antitumor activity, for example, by introducing genes encoding various pro-inflammatory cytokines ex vivo or in vivo.
- Adoptive immunotherapy where ex vivo activated lymphoid cells with defined antitumor reactivity are infused into the tumor-bearing host for treatment.
- Blocking the immune evasion mechanisms of tumors.
- Suppressing immunosuppressive gene expression.

Some of these strategies are nonspecific and aimed at providing a co-stimulatory or an inflammatory signal that can enhance the immune response against tumors while others are intended to present specific tumor-associated antigens to the immune system to generate a desired response against these antigens (El-Aneed 2004; Parney and Chang 2003; Wysocki, Mackiewicz-Wysocka, and Mackiewicz 2002). Some of the important immuno-oncology treatment strategies involving gene therapy are discussed below:

6.2 EX VIVO IMMUNOGENETIC MODIFICATION

6.2.1 ADOPTIVE CELL THERAPY (ACT)

ACT refers to the infusion of lymphocytes to the cancer-bearing host to mediate an effector function. Antitumor lymphocytes capable of high-avidity recognition of the tumor are grown in large numbers (≈ 1011) in vitro and administered to the patient. In ACT, the in vitro activation of lymphocytes provides protection from the in vivo inhibitory factors and enables the manipulation of the host to develop a favorable microenvironment before cell transfer. The administered cells proliferate in vivo and maintain their antitumor functionality, thus making ACT a "living" treatment. The major obstacles with this strategy include the identification of cells capable of selectively targeting antigens expressed on the cancer cells and not on normal cells as well as retrieving these cells from the patient for ex vivo manipulations. The main sources of lymphoid cells include peripheral blood, where immune cells are found to be at very low frequencies; or from within the tumor or tumor-draining lymph nodes. Further, adoptive transfer is technically demanding, patient-specific, time-consuming, and very expensive.

There are three types of ACT that are being evaluated for cancer therapy; these involve (1) natural host cells that exhibit antitumor reactivity called the tumor-infiltrating lymphocytes (TILs), (2) host cells genetically engineered with chimeric antigen receptor (CAR), and (3) T cell receptor (TCR) engineered cells (Fig. 6.2) (June, Riddell, and Schumacher 2015; Kalos and June 2013; Rosenberg and Restifo 2015; Yang and Rosenberg 2016).

ACT employing autologous tumor-infiltrating lymphocytes (TILs) has consistently exhibited clinical efficacy in metastatic melanoma. A substantial improvement in clinical responses was seen when patients received a lymphodepleting preparative regimen before the cell infusion, which induced the production of T cell growth factor, IL-15 (Besser et al., 2013; Rosenberg et al., 2011). Although lymphocyte cultures could be grown from many tumor histologies, melanoma appeared to be the only cancer that reproducibly gave rise to TIL cultures with specific antitumor recognition. Hence, to widen the application of ACT to other human cancers, genetic engineering techniques were utilized to introduce antitumor receptors in normal T cells.

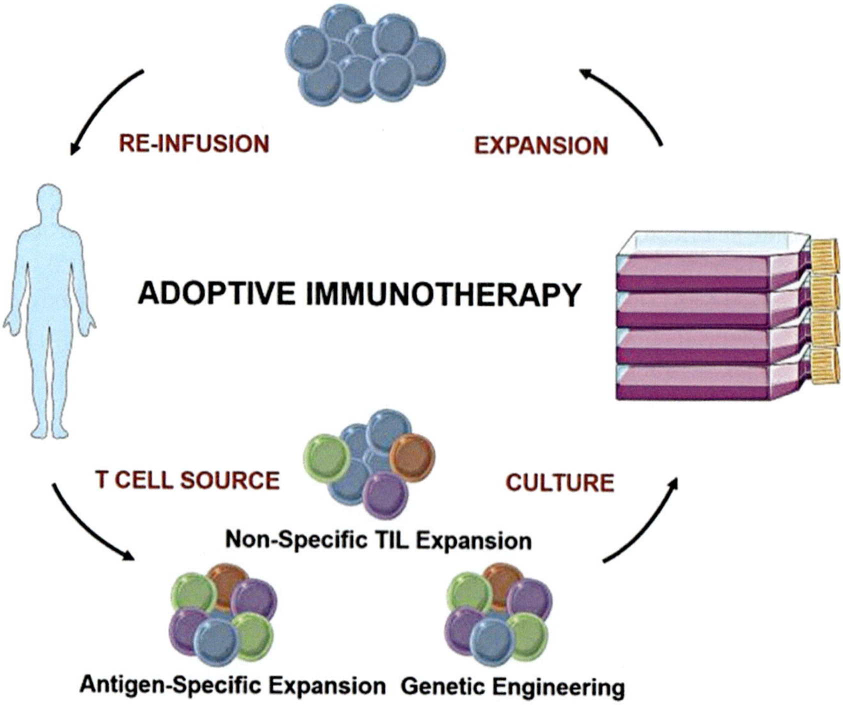

FIGURE 6.2 Schematic diagram of adoptive cell therapy. (Reprinted from Perica, K.; Varela, J. C.; Oelke, M.; Schneck, J. Adoptive T Cell Immunotherapy for Cancer. *Rambam Maimonides Med. J.* **2015**, *6*(1), e0004 under the terms of the Creative Commons Attribution License (http://creativecommons.org/licenses/by/3.0)).

6.2.1.1 GENETICALLY ENGINEERED LYMPHOCYTES FOR USE IN ACT

The specificity of T cells is modulated by integrating genes encoding either conventional α-β TCRs or CARs.

TCRs consist of α and β-chains noncovalently associated with the CD3 complex on the T cell surface and recognize peptides noncovalently bound to major histocompatibility complex (MHC) on the surface of

APCs or tumor cells leading to the activation of T cells (Harris and Kranz 2016; June, Riddell, and Schumacher 2015). TCRs targeting a human lymphocyte antigen A2 (HLA-A2)–restricted peptide from a melanocytic differentiation antigen and MART-1 (melanoma antigen recognized by T-cells 1) epitope were among the initial TCR-based ACT developed against metastatic melanoma. However, normal melanocytes in the skin, eye, and cochlea were affected indicating on-target, off-tumor toxicity (Morgan et al., 2006; Johnson et al., 2009). Fatal neurotoxicity and cardiotoxicity were also reported with two other TCR-based ACT directed to the cancer-testis antigen MAGE-A3 (Linette et al., 2013; Morgan et al., 2013). However, another study involving MAGE A4/wild type TCR gene-modified lymphocytes designed to treat esophageal cancer did not show any toxicities and have been relatively successful with three long survivors (Kageyama et al., 2015). Recently, T cells engineered with TCR specific for HLA-A2-restricted peptide to target the cancer-testis antigen NY-ESO-1 exhibited clinical efficacy without appreciable toxicity and are now under evaluation in a late-stage clinical trial. To improve the efficacy of TCR T cell therapies, TCRs specific to particular tumor neoantigens that are safer than those targeting shared antigens are being developed. Although affinity-enhanced αβ TCR-based engineering approaches have enabled the targeting of essentially all cellular proteins, this approach still suffers from the limitation that it remains susceptible to the common tumor escape mechanisms of MHC down-modulation and altered peptide processing.

Some of the limitations of engineered TCRs such as the need for MHC expression, MHC identity, and co-stimulation can be overcome by employing CARs. They are synthetic polypeptides that contain an extracellular target-binding module isolated from antibodies, a transmembrane module that anchors the molecule into the cell membrane, and an intracellular signaling module. T cells expressing CAR can recognize a wide range of cell surface antigens, including glycolipids, carbohydrates, and proteins. The antigen-ligand engagement in CAR T cells is followed by the production of cytokines, elimination of targeted cells, and the proliferation of T cells, leading to a highly amplified response that can eradicate a huge quantity of tumor cells within weeks. Viral approaches using gammaretroviruses and lentiviruses as well as nonviral approaches such as transposon–transposase systems and gene editing technologies have been explored to introduce CARs into lymphocytes (Curran and Brentjens 2015; Ye et al. 2018).

Although, the initial clinical trials using first-generation CAR designs were disappointing, the first successful clinical application of anti-CD19 CAR gene therapy in humans was reported in 2010 in a patient with refractory lymphoma. The effectiveness of ACT targeting CD19 has been further established in patients with follicular lymphoma, large-cell lymphomas, chronic as well as acute lymphocytic leukemias. In 2017, the US Food and Drug Administration (FDA) approved two CD19-directed CAR T cell therapy-based drugs namely, tisagenlecleucel (Kymriah, CTL019, Novartis) and axicabtagene ciloleucel (Yescarta, KTE-C19, Gilead/Kite Pharma) for the treatment of relapsing/refractory acute lymphoblastic leukemia and non-Hodgkin's lymphomas, respectively (Zheng, Kros, and Li 2018; Pehlivan, Duncan, and Lee 2018). Early-phase clinical trials of CAR T cells targeting BCMA and CD22 in multiple myeloma and acute lymphoblastic leukemia, respectively, have also exhibited favorable results. However, attempts to target tumor-associated antigens in solid tumors have achieved limited success so far. Further, the tumor microenvironment presents additional barriers in the form of immune checkpoints like PD-L1, alterations in the tumor metabolic environment, regulatory T cells, and suppressive myeloid cells. Another alarming issue observed is the toxicity associated with B cell-directed CAR T cells in terms of B cell aplasia and cytokine release syndrome (Sun et al. 2018). Hence, clinical trials that combine CD19-specific CAR T cell therapies with checkpoint inhibitors, switch receptors, suicide gene therapy, and genome editing to address toxicity are under way (Lim and June 2017; Klebanoff, Rosenberg, and Restifo 2016).

ACT strategies have mainly focused on autologous T cells rather than allogeneic donors owing to the inherent barriers imposed by the MHC. If such barriers could be eliminated, universal CAR (UniCAR) T cells derived from healthy donors may be used to overcome many cancer immune defects. This might reduce the treatment costs and further simplify the manufacturing of engineered cells and facilitate the development of "off-the-shelf" ACT products (Torikai et al. 2012; Zakrzewski et al. 2008). Further, efficient genome editing might pave the way for improving the efficacy and safety of ACT. Zinc-finger nucleases, transcription activator-like effector nucleases (TALENS) that rely on customized DNA binding proteins, and the natural bacterial CRISPR-Cas9 system of RNA-guided nucleases, have been employed to introduce DNA double-strand breaks at specific sites to disrupt a gene sequence or provide a site for targeted gene insertion. They can be used to disrupt endogenous TCR genes, selectively edit PD-1 or CTLA-4 genes, or modify T cells to function in the immunosuppressive tumor microenvironment (In-Young Jung and and Jungmin Lee 2018). Thus, although there are several scientific, regulatory, and economic challenges in the use of ACT, an active collaboration from academia and biotechnology and pharmaceutical industry should accelerate the establishment of ACT as a viable approach for common human malignancies.

6.2.2 AUTOLOGOUS TUMOR VACCINES

The earliest attempts at cancer immunogene therapy involved surgically removing tumor cells from the patient, growing them in tissue culture, and introducing immunostimulatory genes using viral vectors in vitro. These cells, when reinjected into the patient, are expected to induce a significant systemic immune response that can eliminate tumor cells and vaccinate the patient against tumor recurrence. Autologous or allogeneic tumor cells genetically modified with the genes that encode cytokines like interleukin (IL)-I, IL-2, IL-4, IL-6, IL-12, granulocyte-macrophage colony stimulating factor (GM-CSF) or γ-IFN have been investigated for their antitumor efficacy in melanoma, colorectal, renal cell carcinoma, neuroblastoma, breast cancer cells, etc. These cytokines modulate the tumor microenvironment, elicit a danger signal that activates APCs like DCs, and induce a strong immune response involving the functional Th1 cells (Bowman et al., 1998; Cayeux et al., 1997; Orengo et al., 2003). The cells expressing GM-CSF were the most potent, long-lasting, and exhibited specific antitumor immunity. This strategy resulted in inhibition of tumor growth and elimination of preexisting tumors in in vitro models (Nagai et al., 1998). However, a pilot clinical trial involving autologous tumor cells expressing B7.2 and GM-CSF as a vaccine did not show encouraging results (Parney et al., 2006).

Another approach involves converting tumor cells into APCs expressing peptides belonging to the MHC class I and class II molecules that facilitate antigen recognition by cytotoxic and Th cells, respectively. Further, introduction of genes encoding co-stimulatory molecules such as B7.1 or B7.2 resulted in tumor cells that are significantly more immunogenic than parental tumor cells and elucidated a strong antitumor immune response mediated by CD8+ cells (Fishman et al., 2008).

Another strategy is to insert an antisense gene in tumors producing high levels of insulin-like growth factor-1 and block its production. This allows immunological rejection of the reimplanted genetically altered tumor and further destruction mediated by cytotoxic T-lymphocytes. This approach has been approved for trials in treating glioblastoma (Trojan et al., 2012).

Whole tumor cell-based vaccines modified to produce GM-CSF (e.g., GVAX) were then investigated and found to be well-tolerated and efficient in a range of cancers including melanoma, prostate, pancreatic, and lung cancers. However, the need for genetic transduction of individual tumor cells proved to be a major limitation of this approach. This was overcome by developing a "bystander" GVAX platform composed of autologous tumor cells and an allogeneic GM-CSF-secreting cell line (John Nemunaitis 2005; Jackie Nemunaitis and Nemunaitis 2003). However, a Phase I/II clinical trial involving patients with advanced-stage non-small cell lung cancer did not show any promising tumor responses (J Nemunaitis et al., 2006).

Recently, tumor cells infected with recombinant viral vectors like poxvirus, vaccinia virus, or nonreplicating fowl pox virus containing multiple co-stimulatory molecules were developed as a vaccine. This approach enhances tumor antigenicity and subsequent T cell response. The Prostvac-VF vaccine (Bavarian Nordic, Inc.) employing a recombinant vaccinia vector as a primary vaccination, followed by multiple booster vaccinations employing a recombinant fowl pox vector, however, failed at Phase III clinical trials (DiPaola et al., 2006). Overall, despite encouraging preliminary results, autologous tumor vaccines have not yet been developed successfully.

6.2.3 DC VACCINES

DCs are the most powerful APCs and the best equipped for antigen identification, initiation of naive T cell responses, T cell co-stimulation, and cytokine production by T-lymphocytes (Guermonprez et al., 2002; Itano and Jenkins 2003). Desired subsets of DCs can be generated from peripheral blood via leukapheresis followed by elutriation and further magnetic bead-based purification. Immature DC cells are developed when cultured in the presence of growth factors and differentiation factors. Cells are then loaded with antigens (peptides, proteins, nucleic acids, and cells) and matured by exposing to cytokines, growth factors, or toll-like receptor (TLR) ligands. The mature DCs are incubated with certain peptides, proteins, or irradiated tumor cells and injected via many different routes of administration. Figure 6.3 describes the key steps in the preparation of DC-based cancer vaccines. Several viral (pox viruses, lentiviruses, adenoviruses, etc.) and nonviral gene transfer vectors (liposomes, cationic polymer complexes, dendrimers, etc.) have been used to load antigens to DCs. DC vaccines have a favorable safety profile, with limited side effects like local inflammatory reaction, flu-like symptoms, or vitiligo-like skin changes (Romani et al., 1994; Butterfield 2013; Chen et al., 2010; Tsao H et al., 2002).

6.2.3.1 GENERATIONS OF DC VACCINES

Between 1995 and 2004, initial clinical trials of early DC vaccines consisting of immature patient-isolated natural DCs or ex vivo-generated monocyte-derived DCs cultured in GM-CSF and IL-4 were conducted. They were loaded with recombinant/synthetic antigenic peptides or tumor cell lysates. In 2010, the US FDA approved the only cellular vaccination product so far, Sipuleucel-T (Provenge) for use in prostate cancer patients (Kantoff et al., 2010). Despite the limited success of DC vaccines against melanoma, non-Hodgkin's lymphoma, and prostrate cancer, owing to the immature status of DCs, the safety and feasibility of DC vaccines were established (Butterfield 2013; Nestle et al., 1998; Hsu et al., 1996).

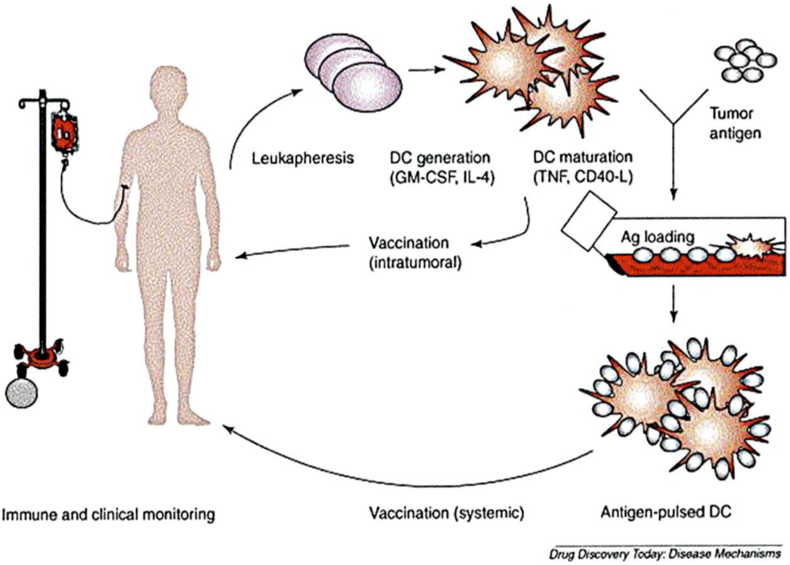

FIGURE 6.3 A schematic diagram describing the key steps in preparation and administration of DCs vaccines for cancer therapy. (Reprinted with permission from Zhang, C.; Engleman, E. G. Mechanisms of Action of Dendritic Cell Vaccines for the Treatment of Cancer. *Drug Discov. Today Dis. Mech.* **2006,** *3*(2), 213–218.)

Second-generation DC vaccines used fully matured monocyte-derived DCs treated with maturation cocktails and mostly employed antigenic peptides derived from tumor antigens such as melanoma differentiation antigens, Wilms tumor 1, and NY-ESO-1. Further, RNA/DNA transfection and cancer cell-DCs fusions were also tested for antigen loading. The clinical trials using second-generation DC vaccines have better results than the first generation confirming the role of maturation signals and tumor antigen sources in the success of DC vaccines (Draube et al., 2011; Wilgenhof et al., 2016; Bloy et al., 2014).

The current research in DC vaccines utilizes specific subsets of DCs equipped with better functionality and reduction in culturing time and costs. For instance, plasmacytoid DCs and myeloid DCs offered better IFN responses and antigen presentation, respectively, and exhibited highly potent T-lymphocyte expansion and memory T cells' generation (Schreibelt et al., 2010; Cella et al., 1999). Further, DCs loaded with tumor-specific neoantigens showed some promising results in melanoma patients. Potential drawbacks of this strategy could be the identification of neoantigens and patient-to-patient heterogeneity, making broad vaccine production difficult. Another strategy would to combine DC vaccines with conventional therapies (chemotherapy/radiotherapy/targeted therapies), immune checkpoints inhibitors, or CAR T cells/TILs-based ACT. Analysis of critical factors like the number of DCs, optimal route of vaccination, number of vaccinations, and level of antigen expression is necessary to improve the efficacy of DC vaccines (Carreno et al., 2015; Santos and Butterfield 2018; Garg et al., 2017).

6.2.4 NK CELL VACCINES

NK cells are highly cytotoxic immune effectors capable of recognizing and killing tumor cells without the requirement of prior antigen exposure. They also act as "helpers" to promote DC and T cell interactions, and eliminate tumors. NK cell malfunctions have been evidenced in many cancers and are thought to aid in the escape of immune surveillance. Hence, it is expected that manipulation of NK cells would result in promising clinical outcomes (Caligiuri 2008; Farag and Caligiuri 2006). Cytokine gene transfer (e.g., IL-2, IL-12, and IL-15) and CAR gene modification of NK cells are the main cancer immunogene strategies involving NK cells. By genetically modifying NK cells to produce cytokines, their survival capacity and proliferation increases resulting in enhanced in vivo activation and antitumor activity. Although clinical trials involving cytokine-modified NK cells went well, the efficacy was limited (Miller et al., 2005; Burga et al., 2016).

CAR gene-modified NK cells (CAR-NK) are expected to eliminate the wider population of tumor cells as tumors without CAR-specific antigens also may be recognized by NK cells with their natural NK cell-activating receptors. The major sources of CAR-NKs are NK-92 cell lines and primary NK cells. The possibility of using gene-editing tools like CRISPR-cas9 to modify NK-92 cell line makes it a promising candidate for "off-the-shelf therapeutic". The main studies investigating the efficacy of CAR-NK cells use tumor-specific antigens like CD19, CD20, human epidermal growth factor receptor 2, epidermal growth factor receptor, etc (Romanski et al., 2016; Müller et al., 2008; Han et al., 2015). Another advantage of CAR-NK cells is that they would elicit less side effects, such as graft versus host disease when compared to CAR T cells as the non-hematopoietic tissues such as liver, kidney, muscle, and lung are not affected by donor NK cells. The preclinical studies involving CAR-NKs have so far been promising; however, clinical efficacy remains to be seen. Reducing the expression of inhibitory NK cell receptors via genetic modification is another possible strategy to enhance NK cell antitumor function (Rezvani and Rouce 2015; Liu et al., 2017; Rezvani et al., 2017).

6.3 IN VIVO IMMUNOGENETIC MODULATION

6.3.1 NUCLEIC ACID VACCINES

Nucleic acid vaccines contain antigens encoded by either DNA or RNA. In a DNA vaccine, antigen-encoding gene(s) are inserted into a bacterial plasmid and administered to the host. Upon subcutaneous or intramuscular administration, DNA enters local cells like fibroblasts or myocytes which then produce and secrete the antigens. APCs will process and present the antigenic epitopes on their surface along with MHC molecules and migrate to the lymphoid organs initiating the desired immune response. On the other hand, RNA vaccines consist of mRNA synthesized by in vitro transcription using a bacteriophage RNA polymerase and DNA template encoding the antigen(s) of interest. The mRNA transcripts are translated after internalization into host cells and the resulting antigens are presented to APCs to stimulate an immune response (Rice, Ottensmeier, and Stevenson 2008; Ulmer et al., 2012). The main advantage of nucleic acid vaccines is the possibility of large-scale production and storage as an "off the shelf" reagent. They are considered to be relatively safe and do not cause infection or autoimmune disorders. DNA vaccines require the DNA plasmids to reach the nuclear membrane of the APCs before transcription and translation into the respective cancer antigen. By contrast, RNA vaccines need to enter the cytosol only for translation and have no oncogenic potential. Hence, they are considered superior to DNA vaccines. However, naked mRNA vaccines are limited by the short extracellular half-life of naked mRNA due to rapid degradation and translocation into the cytosol remains to be a limiting step for RNA-based vaccines. Protamine-condensed naked mRNA with improved stability and personalized RNA mutanome vaccine showed promising results in metastatic melanoma (Weide et al., 2009; Sahin et al., 2017). Single antigen plasmid-based vaccines have been tested for patients with prostrate cancer, metastatic breast cancer, and colorectal cancer. However, significant responses were not achieved. Use of multiple antigens plasmid-based vaccines may result in broadly specific, long lasting, immune response. The co-expression of GM-CSF or chemokines to attract DCs, or the co-expression of immunostimulatory molecules like CD40-L or IL-12 may increase the immunostimulation mediated by the nucleic acid vaccines (Ulmer et al., 2012; McNamara, Nair, and Holl 2015; Amer 2014).

6.3.2 GENETICALLY ENGINEERED VIRAL VECTORS

Oncolytic viruses activate a long-term antitumor response by infecting the tumor cell and initiating specific antitumor immunity. Viral infection of the tumor cell causes cell lysis and release of tumor cell antigens initiating a sustained, specific $CD8^+$ T cell-mediated immune response. The most applied viral vectors include adenoviruses, herpes simplex viruses, retroviruses, lentiviruses, self-amplifying ssRNA viruses like alphaviruses, flavi virus, pox virus, etc (Chiocca and Rabkin 2014; Lundstrom 2018). A genetically modified oncolytic herpes simplex virus type-1, Talimogene laherparepvec (T-VEC), approved by the

FDA in October 2015, is currently marketed in the United States and Europe for the treatment of melanoma lesions in the skin and lymph nodes. The virus was modified by removing two genes for neuronal development and the host immunity was induced by adding the gene for the GM-CSF. Better and durable responses with a favorable safety profile were observed in T-VEC-treated patients. Currently, combination treatments of T-VEC with immune checkpoint inhibitors ipilimumab and pembrolizumab are under clinical evaluation (Conry et al., 2018). Some of the clinical trials involving viral vectors are given in Table 6.1.

Genetic engineering also allows expression of desired cytokines to increase immune response or miRNA response elements to improve the selectivity of oncolytic viruses for tumor cells. Adenoviruses have been utilized to express co-stimulatory molecules like CD80 and CD3 that aid in improving the antigen presentation and immunogenicity of tumors. However, the expression of cytokines by systemic administration of viruses caused transient systemic toxicity due to early transfection events before reaching tumors. Currently, researchers are investigating strategies to overcome this issue. For example, a mutant of a rapamycin-binding protein, FK506 was fused to a cytokine of interest. After administering the oncolytic virus encoding either TNF-α or IL-2, and confirming its confinement to transfected tumors, the cytokine expression can be activated by administration of a small molecule drug. Further, the immune response to the virus can be avoided by intratumoral administration or covalent conjugation of the viral coat with PEG. Another approach could be inclusion of additional genes encoding, for example, GM-CSF, anti-CTLA-4 antibodies, or other co-stimulators in oncolytic viruses to enhance the immune responses (Havunen et al., 2017; Bofill-De Ros, Rovira-Rigau, and Fillat 2017; Marino et al., 2017). Overall, viral vectors have proven to be a flexible and promising approach in cancer immunogene therapy owing to the progress made in vector engineering and safe delivery of viral vectors. However, there is no single universal viral vector suitable for the treatment of all disease conditions. Hence, cautious efforts need to be put in to investigate the optimum virus type for a particular cancer, potential synergistic effects of combination treatments, the route of administration, dosage, etc.

TABLE 6.1 Examples of Clinical Trials of Cancer Immunogene Therapy Involving Viral Vectors.

Disease	Viral vector	Response with reference
Multiple cancers	Enadenotucirev HS HF10	Good safety profile without serious adverse events in phase I (Garcia-Carbonero et al., 2017) Good safety and antitumor activity, further employed for combination therapy with anti-CTLA-4 (Eissa et al., 2017)
Glioblastoma	HS G207	Antitumor activity in phase I Patel et al. 2016)
Colorectal cancer	VV NDV	Induction of immune response (Downs-Canner et al., 2016) Prolonged survival of patients in phase II study (Liang et al., 2003)
Pancreatic cancer	PANVAC	Failure in phase III, encouraging results in new phase I trial (Petrulio and Kaufman 2006)
Prostate cancer	NDV-TAA	Improved survival in phase II (Schirrmacher 2005)
Melanoma	NDV CV CV + PMab TVEC – HS	Failed at phase II/III trials (Slovin et al., 2013) Promising antitumor activity (Voit et al., 2003) Overall response rate 60% with 27% of patients having stable disease (Andtbacka et al., 2015) Phase I trial as neoadjuvant therapy with immune checkpoint inhibitors (Silk et al., 2017)
Solid tumors	NDV	Progression-free survival (Pecora et al., 2002)
Bladder cancer	Instiladrin	Well-tolerated and demonstrated promising efficacy in phase II. Phase III trial ongoing (Boorjian et al., 2017)

HS, herpes simplex virus; VV, vaccinia virus; NDV, newcastle disease virus; PANVAC, vaccinia-fowlpox virus; NDV-TAA, newcastle disease virus-tumor associated antigen; CV, coxsackievirus CVA21 strain; PMab, pembrolizumab; TVEC, talimogene laherparepvec; instiladrin, adenovirus serotype 5 vector.

6.3.3 NONVIRAL VECTORS

A plethora of nonviral technologies have been studied to deliver genetic payload either locally or systemically to target tumor tissue. Nonviral vectors aid in overcoming some of the limitations of viral vectors in terms of immunogenicity, pathogenicity, carcinogenicity, and cost of production (Nayerossadat, Maedeh, and Ali 2012). However, there is a striking difference in the gene transfer efficiency of viral and nonviral vectors.

The efficiency of nonviral vectors is found to be 1/10th to 1/1000th of that reported for viral vectors (Hama et al., 2006). Most of the studies on nonviral vectors for cancer immunomodulation are still in preclinical stages with very few being translated into clinical trials. Majority of them consisted of DNA or RNA packed in liposomes, inorganic nanoparticles, or polymer nanocarriers. Only a handful of studies have entered the clinical trial phase and a few examples are listed in Table 6.2.

TABLE 6.2 Examples of Clinical Trials of Cancer Immunogene Therapy Involving Nonviral Vectors.

Disease	Nonviral vector	Response with reference
Melanoma	Viral capsid proteins	Phase II clinical trial—CpG-ODN and melanocyte differentiation antigen delivery resulted in enhanced T cell response in melanoma patients compared to antigen alone (Goldinger et al., 2012)
Melanoma	PLG matrices	Phase I clinical trial—implantable PLG matrices co-delivering GM-CSF, CpG-ODN, and tumor lysate antigen for recruitment of DCs and $CD8^+$ T cell response (Ali, Emerich, et al., 2009)
Metastatic melanoma	Lipovaxin-MM, a multicomponent DC-targeted liposomal vaccine	Phase I trial resulted in a partial response to the vaccine that was well tolerated without any clinically significant toxicity (Gargett et al., 2018)
Melanoma	Lipo-MERIT, a liposomal tetravalent RNA-lipoplex cancer vaccine	Phase I/II clinical trial revealed that the expressed proteins were presented to the immune system to induce $CD8^+$ and $CD4^+$ T cell responses (Jabulowsky et al., 2018)
Relapsed or refractory leukemia	JVRS-100, a cationic liposome-plasmid DNA complex	Phase I dose escalation study in progress ("JVRS-100 for the Treatment of Patients With Relapsed or Refractory Leukemia - Full Text View - ClinicalTrials.Gov" n.d.)

CpG-ODN, oligodeoxynucleotides with unmethylated cytosine and guanine dinucleotides; PLG, polylactide-co-glycolide; GM-CSF, granulocyte-macrophage colony-stimulating factor; DC, dendritic cell.

Some of the encouraging results in the use of nonviral vectors involved liposomes. One of the earlier attempts at in situ gene modification of tumors involved the direct injection of liposomes encapsulating an allogene that encoded HLA-B7, a foreign antigen that induces an immune reaction against the altered tumor cells (Nabel et al., 1993). In a pilot study involving cationic liposomes, the efficiency of in vivo transduction of the human IFNβ gene was investigated in malignant gliomas (Matsumoto et al., 2008). Another phase I trial investigated the immunogenicity and efficacy of a multicomponent DC-targeted liposomal vaccine, Lipovaxin-MM against metastatic melanoma. The vaccine was well tolerated without any clinically significant toxicity and resulted in a partial response (Gargett et al., 2018). A first-in-human phase I/II study of Lipo-MERIT, a tetravalent RNA-lipoplex cancer vaccine was recently concluded. Four naked RNA encoding melanoma-associated antigens were encapsulated in liposomes and taken up by APCs upon intravenous administration. After intracellular translation, the expressed proteins were processed and presented to the immune system to induce $CD8^+$ and $CD4^+$ T cell responses (Jabulowsky et al., 2018). Another recently concluded clinical trial involved a cationic liposome-plasmid DNA complex, JVRS-100 with potential immunostimulating and antineoplastic activities. It is expected that upon systemic administration, the TLR-targeted cationic lipid-DNA complex enter DCs and macrophages; the DNA binds and activates TLRs, generating NK cells and T cell responses. The results of this study are awaited ("JVRS-100 for the Treatment of Patients With Relapsed or Refractory Leukemia - Full Text View - ClinicalTrials.Gov" n.d.). Similarly, another study revealed that high levels of $CD8^+$ T cells and ovalbumin-specific antibodies were produced when C57BL/6 mice were treated with ovalbumin-encoded pDNA complexed with cationic poly-L-lysine-modified polystyrene nanoparticles (Minigo et al., 2007). Polyethylenimine was also utilized to generate siRNA nanocomplexes that target the TLR in DCs and elicit a therapeutic antitumor response (Cubillos-Ruiz et al., 2009). In addition, nanomicellar systems were also investigated in active immunogene therapy. Cationic micelles containing diethylaminoethyl methacrylate and poly(ethylene glycol) methyl ether methacrylate were used for mRNA condensation and intracellular delivery to DCs (Tang et al., 2010).

Another class of widely investigated nonviral vectors involves polymeric nanoparticles. Poly (lactic-co-glycolic acid), PLGA nanoparticles have been investigated for the delivery of siRNAs, adjuvants, antigens, etc. that activate the immune system. Heo et al. developed PLGA nanoparticles incorporating iRNA for immunosuppressive gene and an immune response modifier (imiquimod, R837) for the activation of DCs through the TLR7 receptors. The results revealed appreciable immunomodulation and antigen-specific tumor therapy in mice. The in vivo migration of DCs to the lymph nodes could also be monitored using PLGA nanoparticles containing an antigen ovalbumin and a NIR fluorophore (Heo and Lim 2014). PLGA nanoparticles carrying DNAs were also employed in an attempt to edit immune cells in vivo by targeting T cells to express leukemia-specific CARs (Stephan et al., 2010). PLGA-based cancer vaccines incorporating antigens, immunostimulatory molecules and/or siRNA against DC immunosuppression were also studied to present antigens to DCs, induce immune activation, and prevent impairment of DCs by tumor-induced immunosupression (Ali, Huebsch, et al., 2009; Hamdy et al., 2011).

Another strategy employed is the use of nanoparticles for targeted delivery of adjuvants to APCs. Synthetic oligodeoxynucleotides containing unmethylated cytosine and guanine dinucleotides (CpG ODN) is a commonly employed adjuvant with potent immunostimulatory effects (Jahrsdörfer and Weiner 2008). Gold nanoparticle complexes carrying CpG ODN along with antigens or surface-modified gold nanoparticles were utilized to enhance DCs homing, initiate their maturation followed by stimulation of both innate and adaptive immune responses (Ahmad et al., 2017; Zhou et al., 2016). Complexation of CpG ODNs with cationic lipids or encapsulation within lipid vesicles have also been attempted to enhance their immunopotency or as a combination therapy with conventional chemotherapy (Wilson and Tam 2009). Another interesting clinical trial involved an engineered virus-like-nanoparticle (self-assembled viral capsid proteins) derived from bacteriophage Qbeta loaded with immunostimulatory oligonucleotides and modified with a melanoma-specific peptide. Upon administration, these nanoparticles drain into lymph nodes, are taken by APCs, which mature and present the peptide leading to activation of cytotoxic T-lymphocytes and Th cells (Goldinger et al., 2012).

The fact that most of the rationally designed nanoparticle-based vectors preferentially accumulates in lymphatic organs aid in targeting APCs and further immunomodulation. However, the clinical success of nonviral vector-mediated cancer immunogene therapy is still a distant goal considering the fact that most of the research is limited to preclinical studies. Scalability and reproducible synthesis methods remain to be a limiting step in the clinical translation of these multifunctional nanosystems. In addition, viral nanoparticles or other immune-targeted systems carry the risk of generating autoimmune responses by mimicking foreign peptides that can cross-react with self-epitopes. Further, development of a standard procedure to assess the long-term toxicity of these nonviral vectors is also a prerequisite before initiating clinical applications. The toxicity profiles of immunotherapeutics are different from conventional cytotoxic agents, for example, inflammation-based adverse reactions are more common among patients receiving immunotherapeutic. Therefore, design considerations for cancer immune nanomedicine would require a different set of preclinical and clinical considerations than those for its traditional counterparts. Further, it is postulated that the strength, duration, and memory of the induced immune responses depends on the release characteristics of the payload and kinetics of antigen presentation. For example, a burst release of the encapsulated gene may induce less pronounced immune responses compared with sustained release particles. Another hurdle associated with immunogene therapy is that only a fraction of the patients exhibit a long-term response. Hence, to improve the outcomes, combination of immunogene therapy with molecularly targeted therapeutics is the need of the hour. In designing rational nanosystems for combination therapy, the main challenge would be to understand the compatibility of mechanisms of actions, to modulate the delivery of different specific agents at desired pharmacological targets, and to be stable until the drugs reach the therapeutic target without any loss of potency. In addition, the loading efficiency of individual agents should be >10 mol% to exert an effective immune response. Further, to achieve long-term prevention of relapse, nanoscale platforms should be designed to generate memory T cell responses against multiple tumor-associated antigens. Engineering nanoparticles capable of in situ amplification of tumor-associated antigens is another hurdle to be overcome in achieving a personalized tumor immune response.

6.4 CONCLUSION AND FUTURE PERSPECTIVES

The past decade has witnessed an enormous amount of development in the field of gene therapy as well as immunotherapy. A better understanding of the mechanisms of induction and immunoregulation of cytotoxic T-lymphocytes and the checkpoint inhibitor pathways has opened a wide arena of possibilities for immune-mediated control of tumor growth. Considering there are multiple mechanisms involved in tumor development and immune evasion, the current research in immunogene oncology is focused on targeting multiple pathways and immune cells that aid in immune surveillance rather than relying on a single drug or target. In this regard, combination strategies involving ACT, gene-modified cancer vaccines, or RNA therapeutics along with the traditional cancer therapeutic modalities like chemotherapy and radiotherapy are evolving. The optimization of dose, schedule and duration of such treatments as well as refining the endpoints that accurately represent overall survival benefit is the need of the hour. Another strategy would be to use high-throughput technologies like immune repertoire sequencing to collect more information on the immune receptors and key molecules of individual cancers as well as the genetic makeup of the host to develop a personalized therapeutic regime. There is also the need to identify biomarkers associated with treatment response and toxicities. Some of the immune-enhancement therapies are associated with adverse toxicities and hence the tumor response-to-toxicity profile needs to be defined clearly while designing immunotherapy regime. In addition, a better understanding of immunological landscape of cancer stem cells may lead to novel therapeutic strategies that can counter tumor relapse.

At the same time, exciting advances in the field of gene therapy like genome editing can contribute in overcoming the current limitations associated with CAR-T cell therapy in terms of time, cost, and difficulty in collecting adequate quantity of good quality T cells from patients. A lot of research in this direction is underway using CRISPR/Cas9 technology attributed with highly effective multiplex gene editing, simple to use, and high pliability. It is also possible to design personalized vaccines that deliver gene combinations for proteins directed against multiple targets and mechanisms. The TriMix technology and "Immunalon" are examples of vectors employing multiple gene combinations for immunogene therapy.

Another strategy would be to integrate nanotechnology to enhance the scope of immunogene therapy. In addition to delivery vectors, nanosystems can also be developed as platform technologies for high-throughput screening of neoantigens, T cell enrichment, expansion, etc. In vivo tracing and recording of the real-time response to the immunomodulators during the course of treatment can be made possible by using functional imaging. Nanoparticles may also be engineered to facilitate in situ amplification of tumor-associated antigens or CAR T cells and when used in combination with immunostimulators or immune checkpoint inhibitors may enhance the specificity of tumor response. Such efforts may help in simplifying and establishing a cost-effective treatment regime.

Finally, it is necessary to develop and validate highly informative and relevant "humanized" preclinical models. This would ensure the identification of the most promising immunogene therapies and combination strategies and accurately predict their toxicities before considering clinical translation. Nonetheless, great strides have been made in the field of immunogene oncology and with focused efforts to overcome the current challenges, improved treatment strategies can be developed to reduce the cancer burden.

KEYWORDS

- **immunogene therapy**
- **genetic engineering**
- **T-lymphocytes**
- **cancer vaccines**
- **immunomodulation**

REFERENCES

Ahmad, S.; Zamry, A. A.; Tan, H.-T. T.; Wong, K. K.; Lim, J.; Mohamud, R. Targeting Dendritic Cells through Gold Nanoparticles: A Review on the Cellular Uptake and Subsequent Immunological Properties. *Mol. Immunol.* **2017**, 91, 123–133.

Ali, O. A.; Emerich, D.; Dranoff, G.; Mooney, D. J. In Situ Regulation of DC Subsets and T Cells Mediates Tumor Regression in Mice. *Sci. Transl. Med.* **2009,** 1 (8), 8ra19.

Ali, O. A.; Huebsch, N.; Cao, L.; Dranoff, G.; Mooney, D. J. Infection-Mimicking Materials to Program Dendritic Cells in Situ. *Nat. Mater.* **2009,** 8 (2), 151–158.

Amer, M. H. Gene Therapy for Cancer: Present Status and Future Perspective. *Mol. Cell. Ther.* **2014,** 2, 27.

Andtbacka, R. H.; Curti, B. D.; Hallmeyer, S.; Feng, Z.; Paustian, C.; Bifulco, C.; Fox, B.; Grose, M.; Shafren, D. Phase II Calm Extension Study: Coxsackievirus A21 Delivered Intratumorally to Patients with Advanced Melanoma Induces Immune-Cell Infiltration in the Tumor Microenvironment. *J. Immunother. Cancer* **2015**, 3 (Suppl 2), P343–P343.

Besser, M. J.; Shapira-Frommer, R.; Itzhaki, O.; Treves, A. J.; Zippel, D. B.; Levy, D.; Kubi, A.; Shoshani, N.; Zikich, D.; Ohayon, Y.; et al. Adoptive Transfer of Tumor-Infiltrating Lymphocytes in Patients with Metastatic Melanoma: Intent-to-Treat Analysis and Efficacy after Failure to Prior Immunotherapies. *Clin. Cancer Res.* **2013,** 19 (17), 4792.

Bloy, N.; Pol, J.; Aranda, F.; Eggermont, A.; Cremer, I.; Fridman, W. H.; Fučíková, J.; Galon, J.; Tartour, E.; Spisek, R.; et al. Trial Watch: Dendritic Cell-Based Anticancer Therapy. *OncoImmunology* **2014**, 3 (11), e963424.

Bofill-De Ros, X.; Rovira-Rigau, M.; Fillat, C. Implications of MicroRNAs in Oncolytic Virotherapy. *Front. Oncol*. 2017, 7, 142.

Boorjian, S. A.; Shore, N. D.; Canter, D.; Ogan, K.; Karsh, L. I.; Downs, T.; Gomella, L. G.; Kamat, A. M.; Lotan, Y.; Svatek, R. S.; et al. Intravesical Rad-IFNα/Syn3 for Patients with High-Grade, Bacillus Calmette-Guérin (BCG) Refractory or Relapsed Non-Muscle Invasive Bladder Cancer: A Phase II Randomized Study. *J. Clin. Oncol*. **2017**, 35 (6_suppl), 279–279.

Bowman, L.; Grossmann, M.; Rill, D.; Brown, M.; Zhong, W.; Alexander, B.; Leimig, T.; Coustan-Smith, E.; Campana, D.; Jenkins, J.; et al. IL-2 Adenovector-Transduced Autologous Tumor Cells Induce Antitumor Immune Responses in Patients With Neuroblastoma. *Blood* **1998**, 92 (6), 1941.

Burga, R. A.; Nguyen, T.; Zulovich, J.; Madonna, S.; Ylisastigui, L.; Fernandes, R.; Yvon, E. Improving Efficacy of Cancer Immunotherapy by Genetic Modification of Natural Killer Cells. *Cytotherapy* **2016,** 18 (11), 1410–1421.

Butterfield, L. Dendritic Cells in Cancer Immunotherapy Clinical Trials: Are We Making Progress? Front. Immunol. **2013**, 4, 454.

Caligiuri, M. A. Human Natural Killer Cells. *Blood* **2008,** 112 (3), 461.

Cancer. http://www.who.int/news-room/fact-sheets/detail/cancer (accessed Oct 16, **2018**).

Carreno, B. M.; Magrini, V.; Becker-Hapak, M.; Kaabinejadian, S.; Hundal, J.; Petti, A. A.; Ly, A.; Lie, W.-R.; Hildebrand, W. H.; Mardis, E. R.; et al. A Dendritic Cell Vaccine Increases the Breadth and Diversity of Melanoma Neoantigen-Specific T Cells. *Science* **2015,** 348 (6236), 803.

Cayeux, S.; Richter, G.; Noffz, G.; Dörken, B.; Blankenstein, T. Influence of Gene-Modified (IL-7, IL-4, and B7) Tumor Cell Vaccines on Tumor Antigen Presentation. *J. Immunol.* **1997**, 158 (6), 2834.

Cella, M.; Jarrossay, D.; Facchetti, F.; Alebardi, O.; Nakajima, H.; Lanzavecchia, A.; Colonna, M. Plasmacytoid Monocytes Migrate to Inflamed Lymph Nodes and Produce Large Amounts of Type I Interferon. *Nat. Med.* **1999**, 5, 919.

Chen, Y.-Z.; Yao, X.-L.; Tabata, Y.; Nakagawa, S.; Gao, J.-Q. Gene Carriers and Transfection Systems Used in the Recombination of Dendritic Cells for Effective Cancer Immunotherapy. *Clin. Dev. Immunol*. **2010**, 12.

Chiocca, E. A.; Rabkin, S. D. Oncolytic Viruses and Their Application to Cancer Immunotherapy. *Cancer Immunol. Res.* **2014,** 2 (4), 295.

Conry, R. M.; Westbrook, B.; McKee, S.; Norwood, T. G. Talimogene Laherparepvec: First in Class Oncolytic Virotherapy. *Hum. Vaccines Immunother*. **2018**, 14 (4), 839–846.

Croci, D. O.; Zacarías Fluck, M. F.; Rico, M. J.; Matar, P.; Rabinovich, G. A.; Scharovsky, O. G. Dynamic Cross-Talk between Tumor and Immune Cells in Orchestrating the Immunosuppressive Network at the Tumor Microenvironment. *Cancer Immunol. Immunother.* **2007**, 56 (11), 1687–1700.

Cubillos-Ruiz, J. R.; Engle, X.; Scarlett, U. K.; Martinez, D.; Barber, A.; Elgueta, R.; Wang, L.; Nesbeth, Y.; Durant, Y.; Gewirtz, A. T.; et al. Polyethylenimine-Based SiRNA Nanocomplexes Reprogram Tumor-Associated Dendritic Cells via TLR5 to Elicit Therapeutic Antitumor Immunity. *J. Clin. Invest.* **2009,** 119 (8), 2231–2244.

Curran, K. J.; Brentjens, R. J. Chimeric Antigen Receptor T Cells for Cancer Immunotherapy. *J. Clin. Oncol.* **2015,** 33 (15), 1703–1706.

DiPaola, R.; Plante, M.; Kaufman, H.; Petrylak, D.; Israeli, R.; Lattime, E.; Manson, K.; Schuetz, T. A Phase I Trial of Pox PSA Vaccines (PROSTVAC®-VF) with B7-1, ICAM-1, and LFA-3 Co-Stimulatory Molecules (TRICOMTM) in Patients with Prostate Cancer. *J. Transl. Med.* **2006**, 4 (1), 1.

Downs-Canner, S.; Guo, Z.S.; Ravindranathan, R.; Breitbach, C.J.; O'Malley, M.E.; Jones, H.L.; Moon, A.; McCart, J.A.; Shuai, Y.; Zeh, H.J.; et al. Phase I study of intravenous oncolytic poxvirus (vvDD) in patients with advanced solid cancers. *Mol. Ther.* **2016,** 24, 1492–1501.

Drake, C. G.; Jaffee, E.; Pardoll, D. M. Mechanisms of Immune Evasion by Tumors. *Adv. Immunol.; Academic Press,* **2006;** Vol. 90, pp 51–81.

Draube, A.; Klein-González, N.; Mattheus, S.; Brillant, C.; Hellmich, M.; Engert, A.; von Bergwelt-Baildon, M. Dendritic Cell Based Tumor Vaccination in Prostate and Renal Cell Cancer: A Systematic Review and Meta-Analysis. *Plos One* **2011**, 6 (4), e18801.

Dunn, G. P.; Old, L. J.; Schreiber, R. D. The Three Es of Cancer Immunoediting. *Annu. Rev. Immunol.* **2004**, 22 (1), 329–360.

Eissa, I.R.; Naoe, Y.; Bustos-Villalobos, I.; Ichinose, T.; Tanaka, M.; Zhiwen, W.; Mukoyama, N.; Morimoto, T.; Miyajima, N.; Hitoki, H.; et al. Genomic signature of the natural oncolytic herpes simplex virus HF10 and its therapeutic role in preclinical and clinical trials. *Front. Oncol.* **2017**, 7, 149.

El-Aneed, A. Current Strategies in Cancer Gene Therapy. *Eur. J. Pharmacol.* **2004,** 498 (1), 1–8.

Farag, S. S.; Caligiuri, M. A. Human Natural Killer Cell Development and Biology. *Blood Rev.* **2006,** 20 (3), 123–137.

Fishman, M.; Hunter, T. B.; Soliman, H.; Thompson, P.; Dunn, M.; Smilee, R.; Farmelo, M. J.; Noyes, D. R.; Mahany, J. J.; Lee, J.-H.; et al. Phase II Trial of B7-1 (CD-86) Transduced, Cultured Autologous Tumor Cell Vaccine Plus Subcutaneous Interleukin-2 for Treatment of Stage IV Renal Cell Carcinoma. *J. Immunother*. **2008,** 31 (1).

Gajewski, T. F.; Schreiber, H.; Fu, Y.-X. Innate and Adaptive Immune Cells in the Tumor Microenvironment. *Nat. Immunol.* **2013**, 14 (10), 1014–1022.

Garcia-Carbonero, R.; Salazar, R.; Duran, I.; Osman-Garcia, I.; Paz-Ares, L.; Bozada, M.J.; Boni, V.; Blanc, C.; Seymour, L.; Beadle, J.; et al. Phase 1 study of intravenous administration of the chimeric enadenotucirev in patients undergoing primary tumor resection. *J. Immunother. Cancer* **2017,** 5, 71

Garg, A. D.; Coulie, P. G.; Van den Eynde, B. J.; Agostinis, P. Integrating Next-Generation Dendritic Cell Vaccines into the Current Cancer Immunotherapy Landscape. *Trends Immunol.* **2017**, 38 (8), 577–593.

Gargett, T.; Abbas, M. N.; Rolan, P.; Price, J. D.; Gosling, K. M.; Ferrante, A.; Ruszkiewicz, A.; Atmosukarto, I. I. C.; Altin, J.; Parish, C. R.; et al. Phase I Trial of Lipovaxin-MM, a Novel Dendritic Cell-Targeted Liposomal Vaccine for Malignant Melanoma. *Cancer Immunol. Immunother*. **2018**, 67 (9), 1461–1472.

Goldinger, S. M.; Dummer, R.; Baumgaertner, P.; Mihic-Probst, D.; Schwarz, K.; Hammann-Haenni, A.; Willers, J.; Geldhof, C.; Prior, J. O.; Kündig, T. M.; et al. Nanoparticle Vaccination Combined with TLR-7 and -9 Ligands Triggers Memory and

Effector CD8+ T-Cell Responses in Melanoma Patients. *Eur. J. Immunol.* **2012,** 42 (11), 3049–3061.

Guermonprez, P.; Valladeau, J.; Zitvogel, L.; Théry, C.; Amigorena, S. Antigen Presentation and T Cell Stimulation by Dendritic Cells. *Annu. Rev. Immunol.* **2002,** 20 (1), 621–667.

Hama, S.; Akita, H.; Ito, R.; Mizuguchi, H.; Hayakawa, T.; Harashima, H. Quantitative Comparison of Intracellular Trafficking and Nuclear Transcription between Adenoviral and Lipoplex Systems. *Mol. Ther.* **2006**, 13 (4), 786–794.

Hamdy, S.; Haddadi, A.; Hung, R. W.; Lavasanifar, A. Targeting Dendritic Cells with Nano-Particulate PLGA Cancer Vaccine Formulations. Lymphat. Drug Deliv. *Ther. Imaging Nanotechnol.* **2011,** 63 (10), 943–955.

Han, J.; Chu, J.; Keung Chan, W.; Zhang, J.; Wang, Y.; Cohen, J. B.; Victor, A.; Meisen, W. H.; Kim, S.; Grandi, P.; et al. CAR-Engineered NK Cells Targeting Wild-Type EGFR and EGFRvIII Enhance Killing of Glioblastoma and Patient-Derived Glioblastoma Stem Cells. *Sci. Rep.* **2015,** 5, 11483.

Hanahan, D.; Weinberg, R. A. Hallmarks of Cancer: The Next Generation. *Cell* **2011,** 144 (5), 646–674.

Harris, D. T.; Kranz, D. M. Adoptive T Cell Therapies: A Comparison of T Cell Receptors and Chimeric Antigen Receptors. *Trends Pharmacol. Sci.* **2016,** 37 (3), 220–230.

Havunen, R.; Siurala, M.; Sorsa, S.; Grönberg-Vähä-Koskela, S.; Behr, M.; Tähtinen, S.; Santos, J. M.; Karell, P.; Rusanen, J.; Nettelbeck, D. M.; et al. Oncolytic Adenoviruses Armed with Tumor Necrosis Factor Alpha and Interleukin-2 Enable Successful Adoptive Cell Therapy. *Mol. Ther. Oncolytics* **2017**, 4, 77–86.

Heo, M. B.; Lim, Y. T. Programmed Nanoparticles for Combined Immunomodulation, Antigen Presentation and Tracking of Immunotherapeutic Cells. *Biomaterials* **2014**, 35 (1), 590–600.

Hsu, F. J.; Benike, C.; Fagnoni, F.; Liles, T. M.; Czerwinski, D.; Taidi, B.; Engleman, E. G.; Levy, R. Vaccination of Patients with B–Cell Lymphoma Using Autologous Antigen–Pulsed Dendritic Cells. *Nat. Med.* **1996,** 2, 52.

In-Young Jung; and Jungmin Lee. Unleashing the Therapeutic Potential of CAR-T Cell Therapy Using Gene-Editing Technologies. *Mol Cells* **2018,** 41 (8), 717–723.

Itano, A. A.; Jenkins, M. K. Antigen Presentation to Naive CD4 T Cells in the Lymph Node. *Nat. Immunol.* **2003,** 4, 733.

Jabulowsky, R. A.; Loquai, C.; Mitzel-Rink, H.; Utikal, J.; Gebhardt, C.; Hassel, J. C.; Kaufmann, R.; Pinter, A.; Derhovanessian, E.; Anft, C.; et al. Abstract CT156: A First-in-Human Phase I/II Clinical Trial Assessing Novel MRNA-Lipoplex Nanoparticles Encoding Shared Tumor Antigens for Immunotherapy of Malignant Melanoma. *Cancer Res.* **2018,** 78 (13 Supplement), CT156.

Jahrsdörfer, B.; Weiner, G. J. CpG Oligodeoxynucleotides as Immunotherapy in Cancer. *Update Cancer Ther.* **2008**, 3 (1), 27–32.

Johnson, L. A.; Morgan, R. A.; Dudley, M. E.; Cassard, L.; Yang, J. C.; Hughes, M. S.; Kammula, U. S.; Royal, R. E.; Sherry, R. M.; Wunderlich, J. R.; et al. Gene Therapy with Human and Mouse T-Cell Receptors Mediates Cancer Regression and Targets Normal Tissues Expressing Cognate Antigen. *Blood* **2009**, 114 (3), 535.

June, C. H.; Riddell, S. R.; Schumacher, T. N. Adoptive Cellular Therapy: A Race to the Finish Line. *Sci. Transl. Med.* **2015,** 7 (280), 280-287.

JVRS-100 for the Treatment of Patients With Relapsed or Refractory Leukemia - Full Text View - *ClinicalTrials.gov https://clinicaltrials.gov/ct2/show/NCT00860522* (accessed Oct 22, **2018**).

Kageyama, S.; Ikeda, H.; Miyahara, Y.; Imai, N.; Ishihara, M.; Saito, K.; Sugino, S.; Ueda, S.; Ishikawa, T.; Kokura, S.; et al. Adoptive Transfer of MAGE-A4 T-Cell Receptor Gene-Transduced Lymphocytes in Patients with Recurrent Esophageal Cancer. *Clin. Cancer Res.* **2015,** 21 (10), 2268.

Kalos, M.; June, C. H. Adoptive T Cell Transfer for Cancer Immunotherapy in the Era of Synthetic Biology. *Immunity* **2013**, 39 (1), 49–60.

Kantoff, P. W.; Higano, C. S.; Shore, N. D.; Berger, E. R.; Small, E. J.; Penson, D. F.; Redfern, C. H.; Ferrari, A. C.; Dreicer, R.; Sims, R. B.; et al. Sipuleucel-T Immunotherapy for Castration-Resistant Prostate Cancer. *N. Engl. J. Med.* **2010,** 363 (5), 411–422.

Kim, R.; Emi, M.; Tanabe, K. Cancer Immunoediting from Immune Surveillance to Immune Escape. *Immunology* **2007**, 121 (1), 1–14.

Klebanoff, C. A.; Rosenberg, S. A.; Restifo, N. P. Prospects for Gene-Engineered T Cell Immunotherapy for Solid Cancers. *Nat. Med.* **2016,** 22, 26.

Liang,W.;Wang,H.;Sun,T.M.;Yao,W.Q.;Chen,L.L.;Jin,Y.;Li,C.L.;Meng,F.J.Applicationofa utologous tumor cell vaccine and NDV vaccine in treatment of tumors of digestive tract. *World J. Gastroenterol.* **2003,** 9, 495–498.

Lim, W. A.; June, C. H. The Principles of Engineering Immune Cells to Treat Cancer. *Cell* **2017,** 168 (4), 724–740.

Linette, G. P.; Stadtmauer, E. A.; Maus, M. V.; Rapoport, A. P.; Levine, B. L.; Emery, L.; Litzky, L.; Bagg, A.; Carreno, B. M.; Cimino, P. J.; et al. Cardiovascular Toxicity and Titin Cross-Reactivity of Affinity-Enhanced T Cells in Myeloma and Melanoma. *Blood* **2013,** 122 (6), 863–871.

Liu, D.; Tian, S.; Zhang, K.; Xiong, W.; Lubaki, N. M.; Chen, Z.; Han, W. Chimeric Antigen Receptor (CAR)-Modified Natural Killer Cell-Based Immunotherapy and Immunological Synapse Formation in Cancer and HIV. *Protein Cell* **2017,** 8 (12), 861–877.

Lundstrom, K. Viral Vectors in Gene Therapy. *Diseases* **2018**, 6 (2), 42.

Manidhar, D.M.; Kesharwani, R. K.; Reddy, N. B.; Reddy, C. S.; & Misra, K. Designing, synthesis, and characterization of some novel coumarin derivatives as probable anticancer drugs. *Med. Chem. Res.* **2013,** 22, 4146–4157.

Marino, N.; Illingworth, S.; Kodialbail, P.; Patel, A.; Calderon, H.; Lear, R.; Fisher, K. D.; Champion, B. R.; Brown, A. C. N. Development of a Versatile Oncolytic Virus Platform for Local Intra-Tumoural Expression of Therapeutic Transgenes. *PLOS ONE* **2017**, 12 (5), e0177810.

Matsumoto, K.; Kubo, H.; Murata, H.; Uhara, H.; Takata, M.; Shibata, S.; Yasue, S.; Sakakibara, A.; Tomita, Y.; Kageshita, T.; et al. A Pilot Study of Human Interferon β Gene Therapy for Patients with Advanced Melanoma by in Vivo Transduction Using Cationic Liposomes. Jpn. *J. Clin. Oncol.* **2008**, 38 (12), 849–856.

McNamara, M. A.; Nair, S. K.; Holl, E. K. RNA-Based Vaccines in Cancer Immunotherapy. *J. Immunol. Res.* **2015**, 794528.

Miller, J. S.; Soignier, Y.; Panoskaltsis-Mortari, A.; McNearney, S. A.; Yun, G. H.; Fautsch, S. K.; McKenna, D.; Le, C.; Defor, T. E.; Burns, L. J.; et al. Successful Adoptive Transfer and in Vivo Expansion of Human Haploidentical NK Cells in Patients with Cancer. *Blood* **2005**, 105 (8), 3051.

Minigo, G.; Scholzen, A.; Tang, C. K.; Hanley, J. C.; Kalkanidis, M.; Pietersz, G. A.; Apostolopoulos, V.; Plebanski, M. Poly-l-Lysine-Coated Nanoparticles: A Potent Delivery System to Enhance DNA Vaccine Efficacy. *Vaccine* **2007**, 25 (7), 1316–1327.

Morgan, R. A.; Chinnasamy, N.; Abate-Daga, D. D.; Gros, A.; Robbins, P. F.; Zheng, Z.; Feldman, S. A.; Yang, J. C.; Sherry, R. M.; Phan, G. Q.; et al. Cancer Regression and Neurologic Toxicity Following Anti-MAGE-A3 TCR Gene Therapy. *J. Immunother. Hagerstown Md.* **2013**, 36 (2), 133–151.

Morgan, R. A.; Dudley, M. E.; Wunderlich, J. R.; Hughes, M. S.; Yang, J. C.; Sherry, R. M.; Royal, R. E.; Topalian, S. L.; Kammula, U. S.; Restifo, N. P.; et al. Cancer Regression in Patients After Transfer of Genetically Engineered Lymphocytes. *Science* **2006,** 314 (5796), 126.

Muenst, S.; Läubli, H.; Soysal, S. D.; Zippelius, A.; Tzankov, A.; Hoeller, S. The Immune System and Cancer Evasion Strategies: Therapeutic Concepts. *J. Intern. Med.* **2016**, 279 (6), 541–562.

Müller, T.; Uherek, C.; Maki, G.; Chow, K. U.; Schimpf, A.; Klingemann, H.-G.; Tonn, T.; Wels, W. S. Expression of a CD20-Specific Chimeric Antigen Receptor Enhances Cytotoxic Activity of NK Cells and Overcomes NK-Resistance of Lymphoma and Leukemia Cells. Cancer Immunol. *Immunother.* **2008,** 57 (3), 411–423.

Nabel, G. J.; Nabel, E. G.; Yang, Z. Y.; Fox, B. A.; Plautz, G. E.; Gao, X.; Huang, L.; Shu, S.; Gordon, D.; Chang, A. E. Direct Gene Transfer with DNA-Liposome Complexes in Melanoma: Expression, Biologic Activity, and Lack of Toxicity in Humans. *Proc. Natl. Acad. Sci. U. S. A.* **1993**, 90 (23), 11307–11311.

Nagai, E.; Ogawa, T.; Kielian, T.; Ikubo, A.; Suzuki, T. Irradiated Tumor Cells Adenovirally Engineered to Secrete Granulocyte/Macrophage-Colony-Stimulating Factor Establish Antitumor Immunity and Eliminate Pre-Existing Tumors in Syngeneic Mice. *Cancer Immunol. Immunother.* **1998,** 47 (2), 72–80.

Nayerossadat, N.; Maedeh, T.; Ali, P. A. Viral and Nonviral Delivery Systems for Gene Delivery. *Adv. Biomed. Res.* **2012**, 1, 27.

Nemunaitis, J. Vaccines in Cancer: GVAX®, a GM-CSF Gene Vaccine. Expert Rev. **Vaccines** *2005,* 4 (3), 259–274.

Nemunaitis, J.; Jahan, T.; Ross, H.; Sterman, D.; Richards, D.; Fox, B.; Jablons, D.; Aimi, J.; Lin, A.; Hege, K. Phase 1/2 Trial of Autologous Tumor Mixed with an Allogeneic GVAX® Vaccine in Advanced-Stage Non-Small-Cell Lung Cancer. *Cancer Gene Ther.* **2006**, 13, 555.

Nemunaitis, J.; Nemunaitis, J. Granulocyte-Macrophage Colony-Stimulating Factor Gene–Transfected Autologous Tumor Cell Vaccine: Focus on Non–Small-Cell Lung Cancer. *Clin. Lung Cancer* **2003**, 5 (3), 148–157.

Nestle, F. O.; Alijagic, S.; Gilliet, M.; Sun, Y.; Grabbe, S.; Dummer, R.; Burg, G.; Schadendorf, D. Vaccination of Melanoma Patients with Peptide- or Tumorlysate-Pulsed Dendritic *Cells. Nat. Med.* **1998,** 4, 328.

Orengo, A. M.; Di Carlo, E.; Comes, A.; Fabbi, M.; Piazza, T.; Cilli, M.; Musiani, P.; Ferrini, S. Tumor Cells Engineered with IL-12 and IL-15 Genes Induce Protective Antibody Responses in Nude Mice. *J. Immunol.* **2003,** 171 (2), 569.

Parney, I. F.; Chang, L.-J. Cancer Immunogene Therapy: A Review. *J. Biomed. Sci.* **2003**, 10 (1), 37–43.

Parney, I. F.; Chang, L.-J.; Farr-Jones, M. A.; Hao, C.; Smylie, M.; Petruk, K. C. Technical Hurdles in a Pilot Clinical Trial of Combined B7-2 and GM-CSF Immunogene Therapy for Glioblastomas and Melanomas. *J. Neurooncol.* **2006**, 78 (1), 71–80.

Patel, D.M.; Foreman, P.M.; Nabors, L.B.; Riley, K.O.; Gillespie, G.Y.; Markert, J.M. Design of a phase I clinical trial to evaluate M032, a genetically engineered HSV-1 expressing IL-12, in patients with recurrent/progressive glioblastoma multiforme, anaplastic astocytoma or gliosarcoma. Hum. *Gene Ther. Clin. Dev.* **2016**, 27, 69–78.

Pecora, A. L.; Rizvi, N.; Cohen, G. I.; Meropol, N. J.; Sterman, D.; Marshall, J. L.; Goldberg, S.; Gross, P.; O'Neil, J. D.; Groene, W. S.; et al. Phase I Trial of Intravenous Administration of PV701, an Oncolytic Virus, in Patients With Advanced Solid Cancers. *J. Clin. Oncol.* **2002,** 20 (9), 2251–2266.

Pehlivan, K. C.; Duncan, B. B.; Lee, D. W. CAR-T Cell Therapy for Acute Lymphoblastic Leukemia: Transforming the Treatment of Relapsed and Refractory Disease. *Curr. Hematol. Malig. Rep.* **2018**, 13 (5), 396–406.

Petrulio, C.A.; Kaufman, H.L. Development of the PANVAC-VF vaccine for pancreatic cancer. *Expert. Rev. Vaccines* **2006**, 5, 9–19.

Kesharwani, R. K.; Srivastava, V.; Singh, P.;. Rizvi, S. I.; Adeppa, K.; & Misra, *K.* A Novel Approach for Overcoming Drug Resistance in Breast Cancer Chemotherapy by Targeting new Synthetic Curcumin Analogues Against Aldehyde Dehydrogenase 1 (ALDH1A1) and Glycogen Synthase Kinase-3 β (GSK-3β). *Appl. Biochem. Biotechnol.* 2015, 176, 1996–2017.

Rezvani, K.; Rouce, R. H. The Application of Natural Killer Cell Immunotherapy for the Treatment of Cancer. *Front. Immunol.* **2015**, 6, 578.

Rezvani, K.; Rouce, R.; Liu, E.; Shpall, E. Engineering Natural Killer Cells for Cancer Immunotherapy. *Mol. Ther.* **2017**, 25 (8), 1769–1781.

Rice, J.; Ottensmeier, C. H.; Stevenson, F. K. DNA Vaccines: Precision Tools for Activating Effective Immunity against Cancer. *Nat. Rev. Cancer* **2008**, 8, 108.

Romani, N.; Gruner, S.; Brang, D.; Kämpgen, E.; Lenz, A.; Trockenbacher, B.; Konwalinka, G.; Fritsch, P. O.; Steinman, R. M.; Schuler, G. Proliferating Dendritic Cell Progenitors in Human Blood. *J. Exp. Med.* **1994,** 180 (1), 83.

Romanski, A.; Uherek, C.; Bug, G.; Seifried, E.; Klingemann, H.; Wels, W. S.; Ottmann, O. G.; Tonn, T. CD19-CAR Engineered NK-92 Cells Are Sufficient to Overcome NK Cell Resistance in B-Cell Malignancies. *J. Cell. Mol. Med.* **2016**, 20 (7), 1287–1294.

Rosenberg, S. A.; Restifo, N. P. Adoptive Cell Transfer as Personalized Immunotherapy for Human Cancer. *Science* **2015**, 348 (6230), 62-68.

Rosenberg, S. A.; Yang, J. C.; Sherry, R. M.; Kammula, U. S.; Hughes, M. S.; Phan, G. Q.; Citrin, D. E.; Restifo, N. P.; Robbins, P. F.; Wunderlich, J. R.; et al. Durable Complete Responses in Heavily Pretreated Patients with Metastatic Melanoma Using T-Cell Transfer Immunotherapy. *Clin. Cancer Res.* **2011**, 17 (13), 4550.

Sahin, U.; Derhovanessian, E.; Miller, M.; Kloke, B.-P.; Simon, P.; Löwer, M.; Bukur, V.; Tadmor, A. D.; Luxemburger, U.; Schrörs, B.; et al. Personalized RNA Mutanome Vaccines Mobilize Poly-Specific Therapeutic Immunity against Cancer. *Nature* **2017**, 547, 222.

Santos, P. M.; Butterfield, L. H. Dendritic Cell–Based Cancer Vaccines. *J. Immunol.* **2018,** 200 (2), 443.

Schirrmacher,V.Clinical trials of antitumor vaccination with an autologous tumor cell vaccine modified by virus infection: Improvement of patient survival based on improved antitumor immune memory. Cancer Immunol. *Immunother*. **2005,** 54, 587–598.

Schreibelt, G.; Tel, J.; Sliepen, K. H. E. W. J.; Benitez-Ribas, D.; Figdor, C. G.; Adema, G. J.; de Vries, I. J. M. Toll-like Receptor Expression and Function in Human Dendritic Cell Subsets: Implications for Dendritic Cell-Based Anti-Cancer Immunotherapy. Cancer Immunol. *Immunother*. **2010**, 59 (10), 1573–1582.

Silk, A. W.; Kaufman, H.; Gabrail, N.; Mehnert, J.; Bryan, J.; Norrell, J.; Medina, D.; Bommareddy, P.; Shafren, D.; Grose, M.; et al. Abstract CT026: Phase 1b Study of Intratumoral Coxsackievirus A21 CVA21 and systemic pembrolizumab advanced Melanoma Patients: Interim Results of the CAPRA *Clinical Trial. Cancer Res.* **2017,** 77 (13 Supplement), CT026.

Slovin, S. F.; Kehoe, M.; Durso, R.; Fernandez, C.; Olson, W.; Gao, J. P.; Israel, R.; Scher, H. I.; Morris, S. Aphase I dose escalation trial of vaccine replicon particles (VRP) expressing prostate-specific membrane antigen (PSMA) in subjects with prostate cancer. *Vaccine 2013*, 31, 943–949.

Spranger, S.; Gajewski, T. F. Mechanisms of Tumor Cell–Intrinsic Immune Evasion. *Annu. Rev. Cancer Biol.* **2018,** 2 (1), 213–228.

Stephan, M. T.; Moon, J. J.; Um, S. H.; Bershteyn, A.; Irvine, D. J. Therapeutic Cell Engineering with Surface-Conjugated Synthetic Nanoparticles. *Nat. Med.* **2010**, 16, 1035.

Sun, S.; Hao, H.; Yang, G.; Zhang, Y.; Fu, Y. Immunotherapy with CAR-Modified T Cells: Toxicities and Overcoming Strategies. *J. Immunol. Res*. **2018,** 2018, 10.

Tang, R.; Palumbo, R. N.; Nagarajan, L.; Krogstad, E.; Wang, C. Well-Defined Block Copolymers for Gene Delivery to Dendritic Cells: Probing the Effect of Polycation Chain-Length. *J. Controlled Release* **2010**, 142 (2), 229–237.

Torikai, H.; Reik, A.; Liu, P.-Q.; Zhou, Y.; Zhang, L.; Maiti, S.; Huls, H.; Miller, J. C.; Kebriaei, P.; Rabinovitch, B.; et al. A Foundation for Universal T-Cell Based Immunotherapy: T Cells Engineered to Express a CD19-Specific Chimeric-Antigen-Receptor and Eliminate Expression of Endogenous TCR. *Blood* **2012,** 119 (24), 5697.

Trojan, J.; Pan, Y. X.; Wei, M. X.; Ly, A.; Shevelev, A.; Bierwagen, M.; Ardourel, M.-Y.; Trojan, L. A.; Alvarez, A.; Andres, C.; et al. Methodology for Anti-Gene Anti-IGF-I Therapy of Malignant Tumours. *Chemother. Res. Pract*. **2012,** 721873.

Tsao H; Millman P; Linette GP; et al. Hypopigmentation Associated with an Adenovirus-Mediated Gp100/Mart-1–Transduced Dendritic Cell Vaccine for Metastatic Melanoma. *Arch. Dermatol.* **2002**, 138 (6), 799–802.

Ulmer, J. B.; Mason, P. W.; Geall, A.; Mandl, C. W. RNA-Based Vaccines. *Vaccine* **2012**, 30 (30), 4414–4418.

Voit, C.; Kron, M.; Schwürzer-Voit, M.; Sterry, W. Intradermal Injection of Newcastle Disease Virus-Modified Autologous Melanoma Cell Lysate and Interleukin-2 for Adjuvant Treatment of Melanoma Patients with Resectable Stage III Disease. *J. Dtsch. Dermatol. Ges.* **2003,** 1 (2), 120–125.

Weide, B.; Pascolo, S.; Scheel, B.; Derhovanessian, E.; Pflugfelder, A.; Eigentler, T. K.; Pawelec, G.; Hoerr, I.; Rammensee, H.-G.; Garbe, C. Direct Injection of

Protamine-Protected MRNA: Results of a Phase 1/2 Vaccination Trial in Metastatic Melanoma Patients. *J. Immunother*. **2009,** 32 (5).

Wilgenhof, S.; Corthals, J.; Heirman, C.; van Baren, N.; Lucas, S.; Kvistborg, P.; Thielemans, K.; Neyns, B. Phase II Study of Autologous Monocyte-Derived MRNA Electroporated Dendritic Cells (TriMixDC-MEL) Plus Ipilimumab in Patients With Pretreated Advanced Melanoma. *J. Clin. Oncol*. **2016,** 34 (12), 1330–1338.

Wilson, K. D.; Tam, Y. K. Lipid-Based Delivery of CpG Oligodeoxynucleotides for Cancer Immunotherapy. Expert Rev. *Clin. Pharmacol*. **2009,** 2 (2), 181–193.

Wysocki, P. J.; Mackiewicz-Wysocka, M.; Mackiewicz, A. Cancer Gene Therapy – State-of-the-Art. Rep. *Pract. Oncol. Radiother*. **2002,** 7 (4), 149–155.

Yang, J. C.; Rosenberg, S. A. Chapter Seven - Adoptive T-Cell Therapy for Cancer. Adv. Immunol.; Schreiber, R. D., *Ed.; Academic Press,* **2016**, 130, 279–294.

Ye, B.; Stary, C. M.; Li, X.; Gao, Q.; Kang, C.; Xiong, X. Engineering Chimeric Antigen Receptor-T Cells for Cancer Treatment. *Mol. Cancer* **2018**, 17 (1), 32.

Zakrzewski, J. L.; Suh, D.; Markley, J. C.; Smith, O. M.; King, C.; Goldberg, G. L.; Jenq, R.; Holland, A. M.; Grubin, J.; Cabrera-Perez, J.; et al. Tumor Immunotherapy across MHC Barriers Using Allogeneic T-Cell Precursors. *Nat. Biotechnol.* **2008,** 26 (4), 453–461.

Zheng, P.-P.; Kros, J. M.; Li, J. Approved CAR T Cell Therapies: Ice Bucket Challenges on Glaring Safety Risks and Long-Term Impacts. *Drug Discov. Today* **2018**, 23 (6), 1175–1182.

Zhou, Q.; Zhang, Y.; Du, J.; Li, Y.; Zhou, Y.; Fu, Q.; Zhang, J.; Wang, X.; Zhan, L. Different-Sized Gold Nanoparticle Activator/Antigen Increases Dendritic Cells Accumulation in Liver-Draining Lymph Nodes and CD8+ T Cell Responses. *ACS Nano* **2016**, 10 (2), 2678–2692.

CHAPTER 7

Gene Therapy for Hemoglobin Disorders

SORA YASRI[1*] and VIROJ WIWANITKIT[2]

[1]*KMT Primary Care Center, Bangkok, Thailand*

[2]*Dr DY Patil University, Pune, India*

[*]*Corresponding author. E-mail: sorayasri@outlook.co.th*

ABSTRACT

Gene is the basic genetic unit of any living things. Several medical disorders of human beings including hemoglobin disorders are related to the underlying gene disorders. The treatment for diseases with underlying gene disorders is usually difficult. To manage with the gene is the new approach in medicine. With advanced medical science approach, the management of the underlying defects of gene becomes the new useful approach. The modern gene related techniques such as gene therapy and gene editing become the new technologies that become the present hope for management of genetic diseases. As a group of genetic disorders, the gene therapy for hemoglobin disorders becomes the new proposed concept for management of those diseases. There are many ongoing researches and trials on gene therapy for management of hemoglobin disorders. Several in silico and in vitro study prove that the gene therapy is useful for management of hemoglobin disorders. In this chapter, the authors discuss the concept and application of gene manipulations for management of important hemoglobin disorders.

7.1 INTRODUCTION

Anatomically, every human consists of several small cell units. In every cell, there is important content inside the genetic material. Gene is the specific thing that regulates the harmonization of everyone's life and controls the inheritance. Gene has been discovered for a very long time. Medical scientists already know that gene is important for health. The phenotypic appearance of any human beings is controlled by gene. Gene is the starting point for signal coding that further finalize in a protein production. The abnormality at any point of the mentioned part can result in the final gross abnormality, the illness. If the problem occurs at gene level, it might result in the aberration of the whole process. The disease that has its rooted cause problem at gene level is medically termed as genetic disease. At present, several diseases are proven for the abnormalities at the gene level and it is accepted that gene defect is an important factor determining the appearance of health problem in the patients.

According to Figure 7.1, the overall process of genetic control of gross function can be seen. Physiologically, the final phenotypic appearance is primarily controlled by the gene or inherited code of everybody. The modification effect of external environment becomes the important factor that results in different expressions. Based on this concept, it can show that the two main factors determining the health and disorder in anybody are gene and environment. Both gene and environment are usually focused in a pathophysiological process of a disease. To successfully manage the disease, the root cause analysis and management of the rooted etiology is needed. However, the management of a disease with underlying gene disorder is usually difficult. To manage the gene is the new concept in medicine. This was not possible in the past. However, with advanced medical science approach, the management of the underlying defects of gene is the possible new useful approach.

Gene manipulation means dealing with the gene. This is the direct involvement of the gene. The expected advantage of gene manipulation is the management of the problem, medical disorder, at its primary origin. At present, the novel manipulation techniques such as gene therapy and gene editing become the new technologies that become the present hope for management of genetic disorders. Several medical genetic medical disorders are studied for the feasibility of using gene management of therapy. As a group of genetic diseases, the gene therapy for hemoglobin disorders becomes the new management concepts. There are many ongoing studies on gene therapy for management of hemoglobin disorders. Many in silico and in vitro study were done and it was proven that the gene therapy might be useful for management of genetic hemoglobin disorders (Fig. 7.2).

Focusing on using genetic manipulation management, the first step is to recognize the underlying genetic problem. The exact genetic aberration must be completely clarified before any further management. The next step is to design and plan for the manipulation tools. The manipulating process might be the new synthesis of genetically active component to replace the ineffective or problematic gene or to trigger the defect genetic portion to regain the normal function (editing). Those designed tools have to be verified in many steps to confirm the effectiveness and safety. The final step is to test in vivo in animal model and human beings. The step on human beings might be far from reality at present but it is the present hope for future success in management of genetic hemoglobin disorders. In this chapter, the authors discuss the concept and application of gene manipulations for management of important genetic hemoglobin disorders.

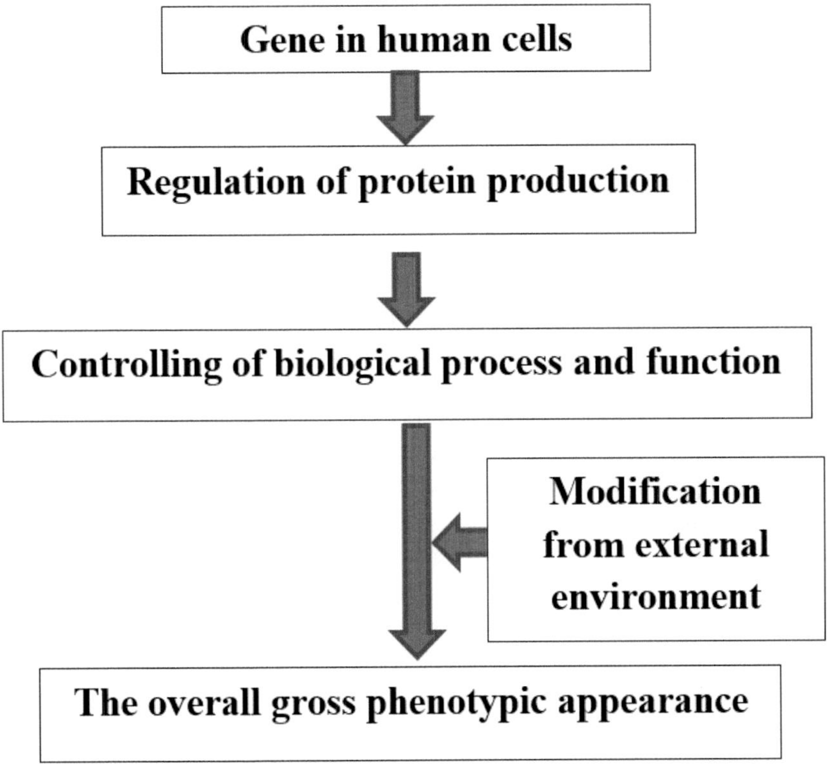

FIGURE 7.1 Brief diagram showing important concepts of gene to phenotype.

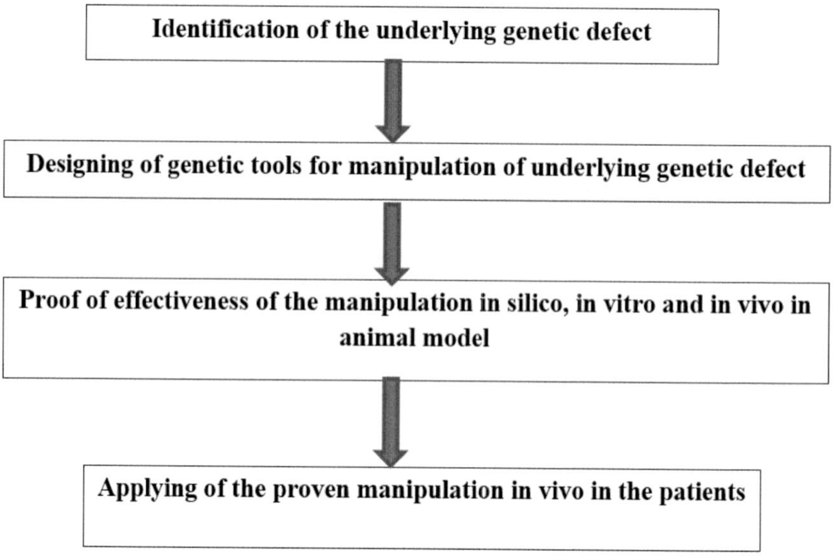

FIGURE 7.2 Brief diagram showing conceptual framework in genetic manipulation of genetic hemoglobin disorders.

7.2 APPLICATION OF GENE THERAPY IN MANAGEMENT OF HEMOGLOBIN DISORDERS

As already mentioned, the gene manipulation becomes an important new medical technology that becomes the hope for management of several medical diseases. In clinical hematology, the applied gene manipulation is usually mentioned for the possible advantages. Since many hematological diseases are proven for underlying genetic etiologies, it is no doubt that gene therapy will be useful in clinical management of those genetic disorders. In brief, hemoglobin disorder is the name of the hematological disease that has the problem of hemoglobin. In case the exact etiology is due to genetic aberration, it is called genetic hemoglobin disorder. The genetic hemoglobin disorder is usually congenital disease. The disease can be seen since the patient is young. The short life span of the patients with genetic hemoglobin disorders can be expected.

At present, genetic hemoglobin disorders are reported worldwide. The disease is common in some specific areas. For example, the hemoglobin S disorder is very common in Africa and the hemoglobin E disorder is extremely high prevalent in Indochina. The affected patients usually have abnormal red blood cells. The clinical problems of anemia and ineffective erythropoiesis are common and these clinical problems can be to morbidity and mortality. In severe case, the affected fetus in utero might abort. However, some kings of hemoglobin disorders are compatible with life but the affected persons will have abnormal growth and development. Genetically, there are two main types of genetic hemoglobin disorder. The first group is namely thalassemia. This kind of genetic hemoglobin disorder has the genetic underlying characteristic of lack of globin chain in hemoglobin due to the genetic mutation. The second group is namely hemoglobinopathy. In hemoglobinopathy, there is still globin chain. Nevertheless, there is an abnormality in hemoglobin chain structure due to mutation that might cause the change of the globin chain structure (malformation, extension, or deletion).

Since the genetic aberration is the main cause of clinical abnormality of genetic hemoglobin disorder, the classical management which is usually the transfusion therapy for correction of anemia is usually not a way to curative management of the disease. The use of gene manipulation on the identified genetic problem might be the new solution for management of genetic hemoglobin disorders. As noted by Chandrakasan and Malik (2014), gene manipulation including gene therapy and gene editing technologies combining with reprogramming somatic cells will be the new therapeutic modality for therapy of genetic hemoglobin disorders. Here, the authors will further summarize and discuss on the applied gene manipulation technology in some important genetic hemoglobin diseases.

7.2.1 THALASSEMIA

Thalassemia is a well-known genetic disease in human beings. This genetic disease results in anemic problem. The genetic abnormality results in disappearance of globin chain component in hemoglobin (Mettananda and Higgs, 2018). The affected globin chain might be beta or alpha chain. If the affected chain is beta globin chain, it is called beta-thalassemia. If the affected chain is alpha chain, it is called alpha-thalassemia. The patients usually have severe anemia and might have abnormal growth. The fragile red blood cells can be seen and there might be the clinical feature of the ineffective erythropoiesis in thalassemia patients (such as short statue, hepatosplenomegaly, and thalassemia face). In many tropical countries, thalassemia is very common. The endemic areas of thalassemia include the Mediterranean and Indochina (Weatherall, 2018). Since the disease is very high prevalence, the disease becomes the main public health problem in the endemic area (Viprakasit and Ekwattanakit, 2018). In Indochina, this disease is extremely common. The local public health policies are to use the antenatal screenings to counteract the problem. The genetic counseling is the main method for prevention of disease. However, due to the failure of complete screening coverage and the controversial issue on using abortion, the disease is still problematic in Indochina. To manage the patients, the classical approach is to use hypertransfusion therapy to correct the anemic problem. Nevertheless, this method is not the curative treatment. In addition, the long-term repeated transfusion can result in several clinical problems such as transfusion-related infection and hemosiderosis (excessive iron storage in the body). The transfusion-dependent thalassemia patients usually die young. The use of gene manipulation for management of thalassemia becomes the new concept for disease management. The details on gene manipulation therapy in different kinds of thalassemia will be further discussed.

7.2.1.1 BETA-THALASSEMIA

Beta-thalassemia is the specific kind of thalassemia where the beta globin chain is affected. The use of gene manipulation therapy for management of beta-thalassemia has been studied for many decades. In 2000, there was a report by May et al. (2000) on using complex lentiviral vectors encoding the human β-globin gene. This became the pioneer approach on using gene manipulation management for beta-thalassemia. After this study, there are continuums of studies. The present approach is the combined usage of lentiviral vector and autologous cell therapy. The safety of using lentiviral vector is the main concern. Bank et al. (2005) showed that the lentiviral vector for beta-thalassemia management was safe and the vector was self-inactivating. The recent trials in few beta-thalassemia patients proved that the lentiviral vector management also significantly reduced the requirement of blood transfusion in studied patients (Sii-Felice et al., 2018; Thompson et al., 2018).

Apart from using lentiviral vector, the new approach based on Bcl11a-mediated γ-globin gene silencing and gene editing is the novel genetic manipulation toward beta-thalassemia. This new approach is aiming at induction of sufficient hemoglobin to solve the problem of blood transfusion independence (Mansilla-Soto et al., 2016). Another similar approach is using CRISPR/Cas9 technology for gamma-globin reactivation (Shariati et al., 2018). These manipulation ways are known as genetic induction. At present, globin gene induction therapy is an important hope for management of beta-thalassemia (Lu, 2014). Finally, there are also attempts to use some approaches that hits at epigenetic level. The examples are the use of JAK2 inhibitors and TGF-β ligand traps that minimize the ineffective erythropoiesis problem (Markis et al., 2016). Based on these evidences, it is concluded that the gene manipulation therapy is the exact hope for curative treatment of beta-thalassemia (Biffi, 2018).

7.2.1.2 ALPHA-THALASSEMIA

Alpha-thalassemia is the specific kind of thalassemia that alpha globin chain is affected (Harteveld and Higgs, 2010). The alpha-thalassemia is usually more severe than beta-thalassemia. Comparing to beta-thalassemia, there are less reports on studies using gene manipulation for management of alpha-thalassemia. Reason is the alpha-thalassemia is usually not compatible with life comparing to beta-thalassemia.

7.2.2 HEMOGLOBINOPATHY

7.2.2.1 HEMOGLOBIN S DISORDER

Hemoglobin S disorder is one of the most well-known hemoglobin disorders. This disorder is highly prevalent in developing African countries. The well-known hematological disorder seen in hemoglobin S disorder is the banana shape red bell cell or sickle cell (Kato et al., 2018). The disease is proven as a genetic adaptation in African patient with a relationship to malaria. There are some reports on using gene manipulation approach for management of hemoglobin S disorder. Wiwanitkit (2009) firstly reported the possibility of using U7snRNA in gene therapy of hemoglobin S disorder. Until present, gene therapy becomes the new hope and is widely studied for clinical application in management of hemoglobin S disorder (Esrick and Bauer, 2018). Since the underlying genetic disorder of hemoglobin S disorder is the mutation on beta-globin gene, the similar conceptual approach as described for beta-thalassemia is widely used. In 2017, there is a publication describing the first trial of hemoglobin S disorder treatment by lentiviral vector-mediated addition of an anti-sickling beta globin gene into autologous hematopoietic stem cells (Ribeil et al., 2017).

7.2.2.2 HEMOGLOBIN E DISORDER

Hemoglobin E disorder is an important hemoglobinopathy. It is extremely prevalent in Indochina. About one-third of the local people in Thailand and Laos are affected by this genetic problem. The patient has the classical feature of genetic hemoglobin disorder. Focusing on pathognomic hematologic feature, the hemoglobin E disease will present a specific abnormal red blood cell that is called target cell. The disease might sometimes be concomitant with thalassemia. The beta-thalassemia hemoglobin E disorder becomes a severe clinical combination. The beta-thalassemia hemoglobin E disorder is the present big public health problem in Indochina. Since hemoglobin E disorder is also the genetic disease involving beta globin gene, the similar concept of gene manipulation for management as described in case of beta-thalassemia can be seen. In 2005, Wiwanitkit (2005) firstly reported the success in management of hemoglobin E disorder using U7snRNA in gene therapy of hemoglobin S disorder. Based on the U7snRNA in gene therapy, some new engineered U7 snRNA is already developed for corrective management of erythropoiesis problem in patients with hemoglobin E disorder (Preedagasamzin et al., 2018). Nevertheless, since this disease is highly prevalent in poor area, Indochina, there are few reports on clinical studies regarding using gene therapy for treatment of hemoglobin E disorder. The most recent interesting report is from Thailand on using one-step genetic correction of hemoglobin E/beta-thalassemia patient-derived induced pluripotent stem cell by the CRISPR/Cas9 system (Wattanapanitch et al., 2018).

7.2.2.3 HEMOGLOBIN C DISORDER

Hemoglobin C disorder is another important hemoglobinopathy. This is also the genetic problem of beta-thalassemia. The first in silico study by Wiwanitkit (2007) also showed the possibility of using gene therapy for management of hemoglobin C disorder. Since this hemoglobin disorder is less common than other famous hemoglobinopathies, the clinical study on this specific kind of hemoglobinopathy is limited.

7.2.2.4 HEMOGLOBIN D DISORDER

Hemoglobin D disorder is another uncommon hemoglobinopathy. This is a report by Wiwanitkit (2013) showing the feasibility of using gene therapy for management of hemoglobin D Punjab disorder.

7.3 CONCLUSION

Hemoglobin disorder is the group of important medical problem seen worldwide. Due to the defect in gene level, the clinical problem of hemoglobin disorder can be seen. The patients mainly have the anemic problems and sometimes require transfusion therapy. To manage the disease, the classical approach cannot result in curative treatment of diseases. Nevertheless, due to the advanced biomedical technology, the new approaches based on gene manipulation become the hope for curative treatment of hemoglobin disorders. There are some reports on the success trail of using gene therapy for management of hemoglobin disorders, both thalassemia and hemoglobinopathy (Table 7.1). It is no doubt that the gene therapy might become the new management for hemoglobin disorders in the near future. Presently, there are several considerations on how to apply the gene therapy for management of hemoglobin disorders. Clarification of genetic etiology of hemoglobin disorder is required and good design and planning for gene manipulation become important processes for successful management of hemoglobin disorders by gene therapy. Also, there must be a good plan for implementation of the new technologies to the real clinical usage in the developing countries where the hemoglobin disorders are usually common.

TABLE 7.1 Current Situations of Using Gene Manipulation in Management of Hemoglobin Disorders.

Diseases	Details
Thalassemia	There are some ongoing researches on using gene therapy for management of some kinds of thalassemia. There are many trials including to human studies regarding using gene manipulation for management of beta-thalassemia. Nevertheless, there is still no standard protocol on gene therapy for management of thalassemia.
Hemoglobinopathy	Similar to general cancer, there are many ongoing researches on using gene therapy for management of hemoglobinopathies. The basic medical researches usually indicate that the gene therapy is possible for curative management of diseases. Many reports are on the use of gene therapy for management of hemoglobinopathies with underlying genetic defect on beta globin gene (such as hemoglobin S disorder and hemoglobin E disorder). Nevertheless, similar to thalassemia, there is still no standard protocol on gene therapy for management of hemoglobinopathy.

KEYWORDS

- **gene therapy**
- **cancer**
- **hemoglobin disorders**
- **drug**
- **thalassemia**

REFERENCES

Bank, A.; Dorazio, R.; Leboulch, P. A Phase I/II Clinical Trial of Beta-Globin Gene Therapy for Beta-Thalassemia. *Ann. N. Y. Acad. Sci.* **2005,** *1054*, 308–316.

Biffi, A. Gene Therapy as a Curative Option for β-Thalassemia. *N. Engl. J. Med.* **2018,** 378, 1551–1552.

Chandrakasan, S.; Malik, P. Gene Therapy for Hemoglobinopathies: The State of the Field and the Future. *Hematol. Oncol. Clin. North Am.* 2014, *28*(2), 199–216.

Esrick, E. B.; Bauer, D. E. Genetic Therapies for Sickle Cell Disease. *Semin. Hematol.* **2018,** *55*, 76–86.

Harteveld, C. L.; Higgs, D. R. Alpha-Thalassaemia. *Orphanet. J. Rare. Dis.* **2010,** *5*, 13.

Kato, G. J.; Piel, F. B.; Reid, C. D.; Gaston, M. H.; Ohene-Frempong, K.; Krishnamurti, L.; Smith, W. R.; Panepinto, J. A.; Weatherall, D. J.; Costa F. F.; Vichinsky, E. P. Sickle Cell Disease. *Nat. Rev. Dis. Primers* **2018,** *4*, 18010.

Lu, Z. M. Globin Gene Induction Therapy for β-Thalassemia. *Zhongguo. Shi. Yan. Xue. Ye. Xue. Za. Zhi.* **2014,** *22*, 237–240.

Mansilla-Soto, J.; Riviere, I.; Boulad F.; Sadelain, M.; Cell and Gene Therapy for the Beta-Thalassemias: Advances and Prospects. *Hum. Gene. Ther.* **2016,** *27*, 295–304.

Makis, A.; Hatzimichael, E.; Papassotiriou, I.; Voskaridou, E. Clinical Trials Update in New Treatments of β-Thalassemia. *Am. J. Hematol.* **2016,** *9*, 1135–1145.

May, C.; Rivella, S.; Callegari, J.; Heller, G.; Gaensler, K. M.; Luzzatto, L.; Sadelain, M.; Therapeutic Haemoglobin Synthesis in Beta-Thalassaemic Mice Expressing Lentivirus-Encoded Human Beta-Globin. *Nature* **2000,** *406*, 82–86.

Mettananda, S.; Higgs, D. R. Molecular Basis and Genetic Modifiers of Thalassemia. *Hematol. Oncol. Clin. North. Am.* **2018,** *32*, 177–191.

Preedagasamzin, S.; Nualkaew, T.; Pongrujikorn, T.; Jinawath, N.; Kole, R.; Fucharoen, S.; Jearawiriyapaisarn, N. L Svasti, S. Engineered U7 snRNA Mediates Sustained Splicing Correction in Erythroid Cells from β-thalassemia/HbE Patients. *Biochem. Biophys. Res. Commun.* **2018,** *499*, 86–92.

Ribeil, J. A.; Hacein-Bey-Abina, S.; Payen, E.; Magnani, A.; et al. Gene Therapy in a Patient with Sickle Cell Disease. *N. Engl. J. Med.* **2017,** *376*, 848–855.

Shariati, L.; Rohani, F.; Heidari Hafshejani, N.; Kouhpayeh, S.; Boshtam, M.; Mirian, M.; Rahimmanesh, I.; Hejazi, Z.; Modarres, M.; Pieper, I. L.; Khanahmad, H. Disruption of SOX6 Gene Using CRISPR/Cas9 Technology for Gamma-Globin Reactivation: An Approach Towards Gene Therapy of β-thalassemia. *J. Cell. Biochem.* **2018**. DOI: 10.1002/jcb.27253. [Epub ahead of print]

Sii-Felice, K.; Giorgi, M.; Leboulch, P.; Payen, E. Hemoglobin Disorders: Lentiviral Gene Therapy in the Starting Blocks to Enter Clinical Practice. *Exp. Hematol.* **2018,** *64*, 12–32.

Thompson, A. A.; Walters, M. C.; Kwiatkowski, J.; et al. Gene Therapy in Patients with Transfusion-Dependent β-Thalassemia. *N. Engl. J. Med.* **2018,** *378*, 1479–1493.

Viprakasit, V.; Ekwattanakit, S. Clinical Classification, Screening and Diagnosis for Thalassemia. *Hematol. Oncol. Clin. North. Am.* **2018,** *32*, 193–211.

Wattanapanitch, M.; Damkham, N.; Potirat, P.; Trakarnsanga, K.; Janan, M.; U-Pratya, Y.; Kheolamai, P.; Klincumhom, N.; Issaragrisil, S. One-Step Genetic Correction of Hemoglobin E/Beta-Thalassemia Patient-Derived iPSCs by the CRISPR/Cas9 System. *Stem Cell Res. Ther.* **2018,** *9*, 46.

Weatherall, D. J. The Evolving Spectrum of the Epidemiology of Thalassemia. *Hematol. Oncol. Clin. North. Am.* **2018,** *32*, 165–175.

Wiwanitkit, V. Usage of U7 snRNA in Gene Therapy of Hemoglobin E Disorder, an in Silico Study. *Gene. Ther. Mol. Biol.* **2005,** *9*, 47–50.

Wiwanitkit, V. Usage of U7 snRNA in Gene Therapy of Hemoglobin C Disorder: Feasibility by Gene Ontology Tool. *Gene. Afr. J. Biotechnol.* **2007,** *6*, 683–684.

Wiwanitkit, V. Usage of U7 snRNA in Gene Therapy of Hemoglobin S Disorder - is it Feasible? *Turk. J. Haematol.* **2009,** *26*, 159–160.

Wiwanitkit, V. Usage of U7 Small Nuclear Ribonucleic Acid in Gene Therapy of Hemoglobin D Punjab Disorder: Rationale? *Indian J. Hum. Genet.* **2013,** *19*, 291–292.

CHAPTER 8

Artificial Intelligence and Biotechnology: The Golden Age of Medical Research

UPENDRA KUMAR[1*] and KAPIL KUMAR GUPTA[2]

[1]*Department of Computer Science and Engineering, Institute of Engineering and Technology, Lucknow 226021, India*

[2]*Department of Computer Science and Engineering, Shri Ramswaroop Memorial University, Barabanki, India*

[*]*Corresponding author. E-mail: upendra48@gmail.com*

ABSTRACT

A significant improvement has been observed in the development of a faster and less invasive diagnostic system for the treatment of diseases by utilizing both artificial intelligence and biotechnology. For example, while for the treatment of cancer, it can be detected at a very early stage, that is, currently still in molecular stages and further a proper treatment can be done before it develops a tumor. The current research in the field of Biotech, Medical Research, and Drug Discovery, though they are high costly affair but has been proved to be highly beneficial for humankind. An extremely popular drug with these current age technology can cure a critical disease for millions of patients across the world and as a result a pharmaceutical company can earn billions of dollars by developing high-tech drugs. Therefore, just to bring one of those blockbuster drugs into the healthcare market, pharma companies spend millions of dollars, if it is not billions in amount and also they spend decades in research, while they do not know where the research will result in fruitful outcomes. Biomedical research landscape is changing and evolving and it has been observed.

Now is the age of genomics and genome-specific drug discovery. Venture firm and medical research firms are investing huge money and resources in genome sequencing and analysis. Undoubtedly, the process of genome sequencing and analysis needs immense computational power; hence, it requires supercomputing approach. Research companies like IBM, NVidia, and other organizations are also attracting high financial investment to build exascale supercomputers to develop better automated diagnostic systems.

8.1 INTRODUCTION

Once famous, Mark Andreessen said, "Software is eating the world." In addition, the computer and hitech industries have had a significant impact on the biotech industry as well as other sectors such as automobile, manufacturing, and many others. It is a combination of tandem software and hardware to enable exascale computing. NVidia CEO, Mr. Jensen Huang said recently in an MIT Technology review that "Software is eating the world, but AI will eat software." In Silicon Valley, technology firms and many stakeholders are now pouring money into artificial intelligence (AI), but where we are now, in most instances, AI means applying AI in the deep learning and medical industries.

Let's make sure the understanding between AI and machine learning's (ML) subtleties. It's not exactly the same stuff. Although the terms AI and ML often use them on an interchangeable basis, ML is a component of AI that gives machines access to data through which machines can learn by themselves.

As far as biomedical research is concerned, data are normally spread around the globe through many different entities. Among other major institutions are universities which conduct primary research, biomedical sample-based biobanks and sample data, pharmaceutical companies that have drug data, and biotechnology companies which hold patient data. It is a complex ecosystem that is spread throughout the world by various entities. Collaboration among these organizations is paramount, including creative relationship models, customer engagement, and confidence in data. It means choosing the right technologies to create the right platform and engagement model. Bioinformatics and big data based in the cloud.

As far as biomedical research is concerned, data are normally spread around the globe through many different entities. Many major institutions include universities conducting primary research data, clinical screening and test data biobanks, drug firms holding medicines and biotech companies holding patient information, and many others. It is a dynamic ecosystem created by various entities around the world. It tends toward the necessity of a flexible and agile platform.

Knowing the significance of computer and profound learning not only includes genome, organ structure, and cell structure, but also patient characteristics, the interaction of drugs with the cells affected, and other external factors. It is not only difficult but almost impossible to train machines based on stand-alone data systems or cloud services. It is possible to create a true platform for ML with the right fabric as the base. Virtualization of information is a popular option for a system that recently began to gain momentum. Integrated Sample Intelligence Data Online Repository (ISIDOR) biostorage project is a great example of a data virtualization architecture built at the heart of the web (Brooks's Life Sciences Blog, 2017).

Big data analytics is also the study of large data sets that include a number of data sets in biomedical research related to healthcare. Big data analytics is aimed at identifying all possible hidden signals, patterns, trends in safety and efficacy. Therefore, the comparisons between potential risk factors and clinical outcomes as well as other valuable scientific data such as the risk or benefit ratio of certain clinical outcomes should be identified. Big data analytical findings could help to evaluate the treatments under examination more effectively or to identify new interventions. It is used to plan future scientific studies to enhance disease control, other clinical benefits, and operations performance.

As researchers generate and use increasingly large, complex, multi-dimensional, and vast data sets, biomedical research quickly becomes data-intensive. But the absence of tools, accessibility, and training often limits the ability to release, locate, integrate, and analyze other generated and data use data. Big data combines knowledge databases from a variety of sources, including archives, randomized or nonrandomized clinical studies, reported or unpublished information, positive or negative clinical outcomes, and database of human services, heterogeneity inside, and over these informational collections will affect the evaluation of treatment impacts of premium. Enormous information investigation gives chances to revealing covered up significant restorative data, deciding potential affiliations or relationships between conceivable hazard factors and clinical results, prescient model structure, approval, and speculation, basic data for arranging of future examinations. To achieve these above-said goals, some statistical methodology and new updated tools need to be developed. Big data analysis provides a lot of benefits; even then there are some issues that need to be addressed like quality, integrity, and validity of the essential information.

As the successful use of big data has grown, so have the expectations of its applications and outcomes in healthcare, such as a significant decrease in-patient mortality, better information regarding patient health and symptoms, reducing readmission, better point of care decision-making, integration of smart devices and sensors with databases, everyday-genome sequencing, developing a treatment approach for cancer, and risk of readmission. While the use of statistics in healthcare excels its use in other industries, the use of big data techniques has lagged in the healthcare. There are six key characteristics of big data that impact its use in healthcare. Volume is one. The massive amounts of data need to be stored somewhere and to be effectively used, it needs to be accessible. Additionally, the data need to be processed before it can be stored. Velocity is another. This refers to how quickly the data are coming which puts a burden on subsequent computing. This is particularly true when the data are captured in real time. Variety is a third. Big data can be textual or numeric and structured or unstructured. In addition to typical information such as patient statistics or epidemic data, big data can include geospatial data, 3D data, audio, video, blog files, and social media. A fourth characteristic and, perhaps, most key is veracity. This refers to the quality of data. Big data can be imprecise, noisy, uncertain, full of biases, and abnormalities. A fifth characteristic is valid. This refers to whether the data are correct and accurate for its intended use. This also goes toward the generalization of the model produced when applied to different populations.

Data mining algorithms can improve the analysis of big data. It is able to extract automatically useful information from a large and complex data set. Advance research techniques are applied to improve performance of extraction of information. Data mining tasks can be subdivided into the following:

- Describing the characteristics and behavior of database.
- Finding patterns which are used for human interpretation.
- Association rule and possible outcomes.
- Predication and mining of data.

There are a lot of data mining algorithms which can be categorized in supervised, unsupervised, and semi-supervised learning.

- Supervised learning means predicting a known target performance using a training set that contains already identified inferences or classifies prospective test data.
- There is no output to forecast in unsupervised learning, but analyzers try to find trends or groupings within unlabeled information that occur naturally.
- Semisupervised learning combine the reliability and accuracy with small data sets and a much wider range of data collection.

The purposes of big data health science are to predict and model the medical data for humanity. Classification, clustering, and regression are common methods used for this purpose. Classification is a supervised example of learning and can be viewed as a simulation model with a categorical output vector or predictive variable. Classification consists of the following basic steps:

- Build rules for assigning objects to a pre-specified class set.
- Variables expected are based on the measurement variable used for these items.
- Traditional statistical approaches, such as systematic regression, Bayesian naive models, decision trees (DTs), neural networks, and support for vector machine, and so on.

To reap significant benefits, integrated big data analytics must be applied to clinical analysis, and clinical integration requires validation of big data analytics' clinical utility. These challenges need to be addressed by stepping up the medical sector's use of big data technology and thus improving patient outcomes and reducing resource waste (Min et al., 2017).

In the global healthcare market, biometrics capturing techniques will continuously improve day by day. It will be driven by different biometric techniques which solve the problem of unauthorized access and increase the customer satisfaction. The improvement and expansion of care delivery over the long term will also be a key objective of these technologies. Biometrics Research Group defines "biometrics of health care" as biometric applications in the offices of doctors, hospitals or for use in patient monitoring. This may include checking access, identifying, managing workforce, or storing patient records. Biomedical security is being deployed in many healthcare institutions and pathology. Secure identification in the healthcare system is critical for both controlling logical access to digitized patients centralized data archives and limiting physical access to buildings and hospital wards, as well as authenticating medical and social support staff.

8.1.1 THE ROLE OF ARTIFICIAL INTELLIGENCE IN BIOTECHNOLOGY AND MEDICAL INDUSTRY

The role of AI in shaping the future of biotechnology and medical industry is significant. At some point in the future, it is not just what we want, but AI is currently applying and we should embrace it. What would become of life if medical advances and improvements were made that promised to extend the quality of life? The role of AI in biotechnology should be looked at and appreciated.

The upcoming future of biotechnology is AI, and it has become a vital part of life science. AI is now advancing into biotech and related industries, but it is not the only industry that is making progress and bringing changes. AI has made some exciting improvements and with many positive results, biotech companies and pharmaceutical companies are using AI in drug discovery (Riordon et al., 2019).

Some of the essential contributions to medical industry so far include

- Monitoring performance of individual patient.
- Aggregating patient treatment by selected characteristics such as major, illness study, medicine performance, and so on.
- Identification and development of effective instruction techniques.
- Standard treatment techniques and instrument analysis.
- Testing and validation of healthcare system.
- Providing valuable feedback.
- Tracking treatments.

8.1.2 MACHINE LEARNING—THE NEXT STEP IN BIOMEDICAL RESEARCH

AI's adoption of ML enables it through its own systematic testing to solve complex problems. Instead of coding what it needs to know, developers create AI capable of learning and analyzing data—resulting in innovative solutions that are virtually unattainable by human ability alone. Over the past decade, unprecedented data sets have become available which can be added and accessed by professionals worldwide. For even the most experienced analysts, global data amalgamation is too vast, but AI programs with ML capabilities can interpret the data into usable solutions.

The technological limitations have held back the biomedical field, but ML and AI programs have broken into new possibilities through the barriers. Trends in security and healthcare suggest that future generations will make daily use of biotech, either to thwart identity theft or to cure cancer. We may still ask the big questions, but the solutions are being found by AI programs.

Public opinion on biomedical research based on AI is mixed. Healthcare has been positively affected by 79% of the public. On the other hand, stem cell research is opposed by 44% of the public. These numbers show a distortion of public knowledge about biomedical industry and at least believe that it should be used in some aspects of healthcare and not others. However, the manipulation of organic materials and the level of microscopy have already produced new drugs and treatments, while ethics committees discern what is acceptable.

Meanwhile, biometrics—cousin of biotechnology focused on individuals, storage, and identification through features and behavioral patterns—also saw benefits from our increased use of data centers and ML. There will be 30 billion internet-connected devices by 2020, and data centers will store their information remotely. With our use of big data, Internet of Things (IoT) and data center utilization growth correlates to reveal patterns and solutions, increasing biometric options and accuracy (Ayers, 2018).

8.1.3 STAKEHOLDERS OF HEALTHCARE BIOINFORMATICS SYSTEM

Stakeholders could also be a private, group, organization or any system that affects or will be full of aid system. Figure 8.1 show the details of stakeholders of healthcare bioinformatics system. Detail description of some stakeholders listed below.

8.1.3.1 PERSONS SUFFERING FROM SERIOUS DISEASE

With customized recommendations, the patient community forever expects to make use of a wide variety of assistance services at a reasonable price (Mancini et al., 2014). In addition to clinical identification by physicians, additional medical data need to be realized through digital platforms such as Facebook, Twitter, clinical forums, and so on. These huge sources of information change patients to join similar individuals to gain data such as symptoms of disease, various effects of chemical drugs, hospitalization, data on drugs, and feedback on clinical reports and postimpact situations with enhanced privacy (Bouhriz and Chaoui, 2015). World Health Organization area unit unable to go to hospitals, associate degree avail telemedicine services for his or her aid desires (World Health Organization, 2009). The system will serve as a massive data repository and capture significant health indicators such as rapid changes in body temperature, heartbeats, blood pressure, and information flow into a centralized repository to cause periodic health warnings.

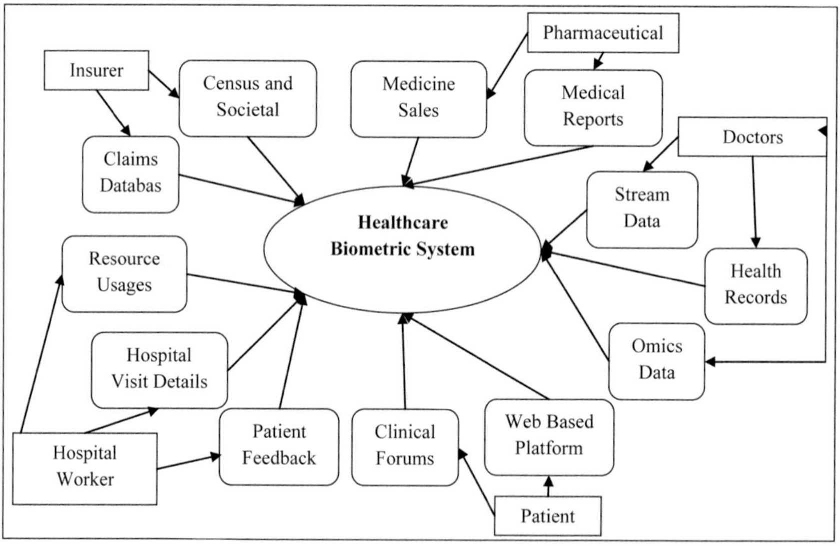

FIGURE 8.1 Stakeholders of healthcare bioinformatics system.

8.1.3.2 DOCTORS FOR THE TREATMENT OF DISEASE

The vast amount of knowledge produced by various phases of patient identification and treatment plans allows suppliers gain insight into the success of the area unit treatments. Throughout the execution of treatment plans, there are several sources of massive information that area unit generated by the aid system. This contains identification codes for various diseases and health facilities, laboratory reports, clinical records, data on medical imaging, and tools for sensing elements that record the behavior of patients in completely different situations. Once such a huge area source of information unit thinks about building the clinical disease repository, it improves public health and provides faster response by analyzing disease patterns effectively. Furthermore, the integration of knowledge from wearable devices provides important advantages, such as facilitating the use of physicians.

8.1.3.3 HOSPITAL WORKERS FOR HANDLING DAY-TO-DAY CARE SERVICES

Hospital operators rely heavily on the outcome of massive sources of information to manage the patient's experiences effectively and to optimize market resources. Models supported the unit of the area of prognostic and prescriptive analytics developed with the knowledge-base for the strength of relationships between indexes of patient satisfaction and used services. Furthermore, resource allocation and optimization techniques will be successfully deployed to meet the staff needs of different sections of the hospital on the idea of accessible huge information. Strategic managers of the hospital can also use the data on situation awareness to agree on the colocation of different departments to optimize the use of high-priced aid instrumentation. Additionally, developing descriptive models on the idea of posttreatment information generated from follow-up phone calls, e-mail communications, text messages would make it easier to boost the services offered.

8.1.3.4 PHARMACIST AND PATHOLOGIST

Massive information's impact reflects the pharmaceutical and clinical research domain reform. Using clinical information helps make prognosticative models for understanding the processes of biology and drugs that attribute to the high rate of success to achieve effective drug senses (Wu et al., 2017). The health information analysis from wide-ranging huge information sources helps pharmaceutical company corporations to live the result of designed medicine with smaller and shorter trials (Merelli et al., 2014). Through integrating computing technology with machine-controlled systems in drug-producing plants, pharmaceutical companies can easily combine and analyze multiple types of information to create end-to-end consumer solutions. Inputs from different stakeholders, such as a doctor's recommendation for a selected disease, amount of patient consumption, history of drug retailers' sales, change the pharmaceutical organizations to measure and visualize their current market position for strategic inbound business choices.

8.1.3.5 HEALTH INSURANCE COMPANIES

As per the increase in healthcare progresses and patients, it is necessary to open new analytical financial revenues for the advantage of insured persons. Consequently, the new health plans for off-time diseases supported by the realms with the lowest premium price will be introduced. Advocating acceptable health plans for buyers on the idea of varied options such as age, gender, case history, income, nature of job allows benefits for each underwriter and customer insurance. Using prognosticative modeling techniques to analyze unstructured information from claim history allows the institutions to predict patterns of true claims and strange outliers to minimize the misuse. IoT analyzes customer behavior and aims to implement new and innovative business models such as insurance based on use of insurers. Mobile IoT plays an important part in the restoration of services by encouraging new business models to evolve and improve business processes, increase efficiency, and customer experience (Dimitrov, 2016). By analyzing the influence of massive information on the higher than stakeholders, in an exceedingly typical patient-centered aid system, we have known the potential huge sources of information. Table 8.1 presents numerous analytical methods that investigate purposeful interest patterns within the aid domain. In the next section, we reviewed numerous aid frameworks by shedding light on their sources of information, analytical capacity, and application areas (Palanisamy and Thirunavukarasu, 2017).

TABLE 8.1 Commercial Drugs Discovered by Machine Learning Tools

Year	Name of drug with their use	Machine learning tool	Developer
2012	CCT244747 used to inhibit checkpoint kinase 1	Docking	The Institute of Cancer Research, United Kingdom
2014	PTC725 used inhibit hepatitis C RNA replication	SAR/QSAR	PTC Therapeutics, Merck Research Laboratories, United States of America
2015	RG7800 used to treats spinal muscular atrophy	SAR/QSAR	Pharma Research & Early Development, PTC Therapeutics and SMA Foundation, United States of America
2016	GDC-0941 used to inhibit phosphatidylinositol-3-kinase	Molecular modeling	The Royal Marsden National Health Service Foundation Trust and The Institute of Cancer Research, United Kingdom

8.1.4 FACTORS INFLUENCING THE IMPLEMENTATION OF MACHINE-LEARNING TECHNIQUES FOR THE IMPROVEMENT OF BIOMEDICAL SYSTEMS

8.1.4.1 ORGANIZATION CAPABILITY

The capability of the organization means the power of managing the machine-learning techniques (MLTs) effectively and is thought to be the key success considers the applying of huge knowledge in the domain (Davenport, 2014; Kung et al., 2015). The things embrace within the organization capability embrace prime management support, MLT strategy, intensity of structure learning, collaboration, information exchange, basic resources like finance, and supply information (Abdullah et al., 2005).

8.1.4.2 TECHNOLOGICAL CAPABILITY

Technological capability refers to the ability to convert comprehensive knowledge into valuable information and to provide useful information to policy makers through IT infrastructures and analytical systems. Scalability, functionality, practicality, and versatility are the characteristics of technological capabilities (Lakulu et al., 2010).

8.1.4.3 HUMAN AND TECHNICAL CAPABILITY

The achievement of MLT depends heavily on the human ability to assess the nature of extensive knowledge. Personnel with technological and administrative skills consisted of people. Technical capabilities are the capacity of trained associates to perform tasks of vast complexity (e.g., someone with analytical abilities or experience) (Davenport, 2012).

8.1.4.4 ANALYTICAL CAPABILITY

Analytically, ML systems are able to collect, store, process, and analyze comprehensive information that is translated into valuable information and that facilitates decision-making in an efficient way. To integrate the ability to provide fast information, traced ability, prophecy, and ability, analysis capabilities are necessary for ML applications (Talib et al., 2011). As regards mass management of the data, the analytical capacity requires sufficient space, preprocessing, and system analyzes (e.g., handling, measuring efficiency, honesty, lack of consistency, improvement of resources, and speed) (Siddiqa, 2016).

8.1.4.5 ANALYTICS CULTURE

The analytics culture refers to the practice of the company to integrate analytics results into the process of decision-making. Extremely addicted to the analytics reports in decision-making, massive knowledge in the activities of the organization has created positive acceptance toward victimization. Analytics community consisted of a philosophy of decision-making for data-driven higher cognitive processes and analytics.

8.1.4.6 SETTING ENVIRONMENT ISSUES

It refers to recognizing the value of vast awareness of victimization in improving the quality of the system. This problem could promote the advancement of the capability of AI and the development of new business model. The issues include organizational understanding of the context of ML, organizational preparation, competitive pressure, and capability.

8.1.4.7 KNOWLEDGE MANAGEMENT

The management of knowledge could be a significant strategy for applying the framework to provide valuable insights on data and decision-making skills. AI's development is focused on ML methods in the medical industry. This is influenced by the effectiveness of tasks such as data administration, the provision of information, access to knowledge, integrated knowledge, the provision of knowledge, and privacy (Halaweh and El Massry, 2015).

8.1.4.8 KNOWLEDGE AND INFORMATION QUALITY

Massive data analytics quality could be a vital part for effective decision-making in the MLT implementation cycle because it is commonly utilized by call manufacturers at intervals the decision-making method (Gorla et al., 2010). Knowledge quality is commonly outlined because the handiness of information that meets user needs. Previous studies recommended that information value be convenient, accurate, available, effective, complete, standard, and accessible (Hou, 2013).

Meanwhile, the value of information refers to the knowledge gained from processed analytics. Important for system efficiency are the implications of quality in data processing analytics. The knowledge quality of the created data included completeness, accuracy, format, and currency (Wixom and Todd, 2005).

8.1.4.9 SYSTEM QUALITY

The quality of the system refers the efficiency of the analytics software systems in the personalized decision-making process of support organizations. An empirical study of an associate degree clearly found that the value of the program had a strong influence on the price of business and firm efficiency. The performance of the system was adapted from a successful model consisting of system consistency, system usability, system efficiency, system time interval, system integration, and system privacy (Petter et al., 2008).

8.2 EXISTING MEDICAL RESEARCH USING ARTIFICIAL INTELLIGENCE AND MACHINE LEARNING TECHNIQUES

8.2.1 DEEP LEARNING IN BIOMEDICAL RESEARCH

Authors in this section describe the deep-learning techniques in biomedicine. ML can take place in one of two ways without explicit programming: conventional learning, or deep learning (nonlinear distributed networks of neurons with many hierarchically dependent layers) with a single hidden layer or vector-supporting machines (Mamoshina et al., 2016). Deep learning has recently been thoroughly reviewed by LeCun et al. (2015). The difference between deep and shallow learning is that shallow learning is not handling the raw data well, it requires extensive human input to be developed and maintained, while in-depth learning is mostly unattended when it is in motion (LeCun et al., 2015). Bengio and LeCun referred to this as maximizing the scope of exchange, which means that only a deep circuit can perform computational tasks that are exponentially complex without needing an infinite number of elements (Greene and Troyanskaya, 2012).

Most deep neural networks (DNNs) can be classified into three major categories (Bengio et al., 2015):

- *Unsupervised learning networks*: The purpose is to capture the correlation of high-order data through jointly identifying statistical distributions with associated classes where possible. The Bayes rule can later be used to create a discriminatory learning machine (Krizhevsky et al., 2012).
- *Supervised learning networks*: Built to provide full discriminative power in classification issues and trained with only marked data (Krizhevsky et al., 2012).
- *Hybrid or semisupervised networks*: The aim is to detect data using the (unsupervised) device inputs. Data are typically used to train weights before the surveillance step to speed up the learning process.

In biological research applications, such as annotation, semantic linking, DNNs have the potential to benefit complex biological data, including biomarkers, drugs discovery, drug repurposing, and clinical orders should be interpreted. DNNs can have a major effect in transcriptomic data analysis. Several million samples of human transcriptomal information are available from several repositories over nearly two decades, including GEO, ArrayExpress, Encyclopedia of DNA Elements (ENCODE), and The Cancer Genome Atlas (TCGA). The broad institute connectivity map and LINCS project cell line data are available as shown in Figure 8.2.

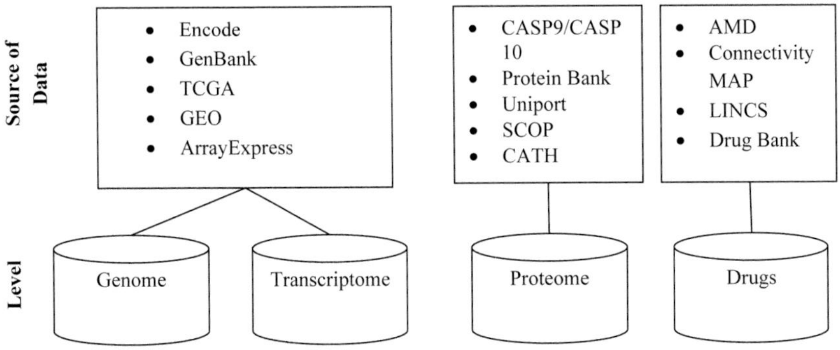

FIGURE 8.2 Potential biological data sources that can be fed into deep neural networks.

8.2.2 LOCAL MESH PATTERNS IN BIOMEDICAL IMAGE INDEXING AND RETRIEVAL

The local binary pattern (LBP) has its own demerits like noise sensitivity in images. Cathode ray local binary pattern (CRLBP) was implemented to resolve this demerit that used local gray weighted rate. It replaces the center pixels' conventional gray quality, but it was not very significant. Completed hybrid LBP that was reliable and stable was proposed to solve this new hybrid system. It used derivatives of the first and second order combined to represent the pattern on images. The average local gray level is adopted instead of traditional level of neighbors as well as center values.

Xiao describes a new way for rotation invariant texture analysis using Rodon and Fourier transformation (Xiao et al., 2007). Zeng described a method for texture representation that was pixel pattern based and independent on variations of illumination and rotation (Zeng, 2017). Singh et al. (2014) also proposed a method of rotation invariant texture analysis using radon and polar complex exponential transform in which rotation invariance is achieved by Rodon Space & Translation was achieved by normalization of moment of Polar Complex Exponential Transform

(PCET). A. Suruliandi will encode patterns that are not uniform in more than one function to set specific characteristics and better classification rates. Implementing the mapping proposed on our data set shows that the proposed method will increase the classification accuracy. The proposed approach also improves the identification levels for all kinds of LBPs, especially those with large communities (Meena and Suruliandi, 2011). By using a combination of Speed Up Robust Transform (SURF) and LBP base on bag of features, Wisarut Surakarin and Prabhas Chongstitvatana (2015) suggested recognition types of clothing. With the height accuracy score, the three subwindows will improve and enhance the quality of interest point detection. The experiment showed that the proposed method achieves a 73.57% accuracy rating. Amir Reza Rezvan Talab and Mohammad Hossein Shakoor have proposed a new mapping technique for LBPs in texture classification. The proposed solution would code in more than one attribute nonuniform patterns to define differentiating characteristics and classification levels. Implementation of the proposed mapping on our built data set shows that the proposed method will increase classification accuracy. The approach suggested also improves identification rates for LBPs of all types, especially large communities (Talab and Shakoor, 2018). LBP and its variants dominant local binary pattern (DLBP), local derivative pattern (LDP), and advanced local binary pattern were investigated by K. Meena and Dr. A. Suruliandi. Facial characteristics are extracted using the nearest K-neighbor classification algorithm and compared. Distance measurement of data is used for identification. Experiments have been performed on the databases of Japanese Female Facial Expression (JAFFE) woman and Yale head. Results show that LDP performs much better consistently than the other methods. The two main demerits of the LBP were analyzed by Jing-Hua Yuan, De-Shuang Huang, Hao-Dong Zhu, and Yong Gan, and a novel order-based center-symmetric local document was proposed. Do not review any of the current designations, please. Binary pattern (OCSLBP) capture more detailed discrimination (Yuan et al., 2014).

Sadat et al. (2011) proposed a method of image recovery. The proposed method comprise a picture definition called half-LBP. The new method of picture definition called half-LBP was introduced by Najmus Sadat and Abdur Rakaib. Half LBP converts the relation between the intensity levels of local pixel blocks into a binary model so that the dimensions of the descriptor are significantly reduced while the accuracy of the recovery is comparable to state-of-the-art methods (Sadat et al., 2011).

8.2.3 HEALTHCARE-AS-A-SERVICE USING FUZZY RULE-BASED MACHINE LEARNING

When information and communication technology expands, remote applications in which patients are treated from abroad are increasing exponentially. Information collected about patients in remote applications varies with size, range, flexibility, and values as a result to huge data. The process, such as the large range of heterogeneous expertise, is one of the most critical challenges that requires a specialized approach. To address the difficulties, the related goal is to provide a modern, fuse rule-based cluster for healthcare-as-a-service (HaaS). The proposed topic is focused on the initial cluster development, recovery, and comprehensive cloud-based data process. A fuzzy rule-based classifier is designed to make economic decisions on information classification within the planned subject. Membership functions are designed to make information from the data collected for fuzzing and defuzzification processes.

Figure 8.3 shows the network model used in the planned system. This system is divided into three layers, in particular the layer of acquisition of knowledge, the layer of transmission and operation. It is the duty of the information received from the collection layer to gather patient information from entirely different geographic areas. This information can be created from the network of body sensors, the ad-hoc transport network and the network of devices used in homes or in many hospitals (Lu et al., 2014). The data on the transportation network are generated by body sensors or by sensors installed within the vehicles by the unit area. Usually this vehicle area unit includes hospital vans and ambulances, but patients must send the information from their personal vehicles if this area unit is equipped with appropriate sensors. The equipment on these networks transfers the data separately or selects a network manager which supervises knowledge transfer for each of the networks. The Internet communication between devices involves the use of short-term networking technologies such as Bluetooth, ZigBee, RFID, broadband (UWB), and 60 gigactic mm wave. Wav. As the Internet works are done, devices, body sensors, and vehicles should form a mobile knowledge transfer cluster head and should be inhumed or intracoalited (Kumar et al., 2013).

Data from the second layer is used for transmission purpose. It is responsible for the process of transmission and communication within and between cloud-connected networks. To this end, it is presumed that completely different gateways are located in various locations such as hospitals, schools, and alternative components of a country. Short-range communication methods such as Dedicated Short-Range Communications (DSRC)/ Wireless Access in a Vehicular Environment (WAVE) (shorter range networking/wireless access) and dynamic spectrum access (Samsung Research America, SRA) are used. The network connectivity system is accessed by long-term network methods, such as WiFi, Worldwide Microwave Access (WiMax), and Future Evolution (LTE) and its version of LTE's advanced area unit (Kumar et al., 2015).

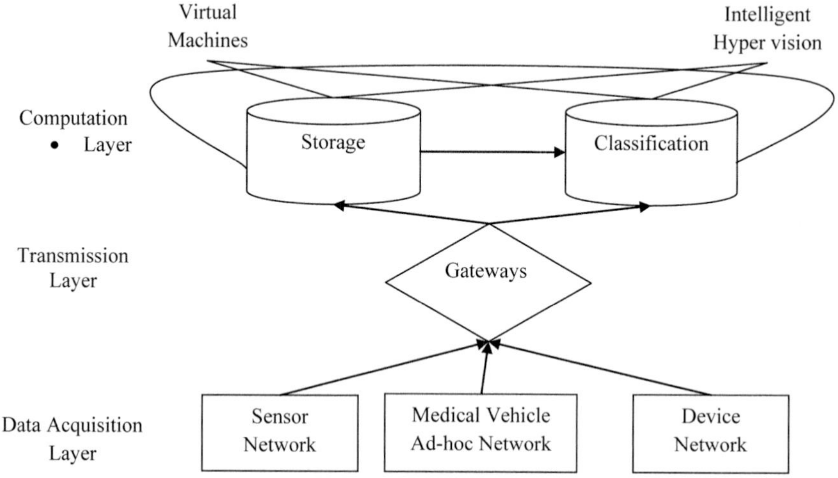

FIGURE 8.3 Networks model for planned healthcare-as-a-service (HaaS).

Third layer is responsible for the computation. All the data produced from the networks or devices passes through the transmission layer of the cloud server whenever the configured database receives the information and stores it in completely different subclouds. Then, the area unit results are determined to meet the demands from completely different buyers. The vector machine programming and allocation of resources is done here with the aid of a smart hypervisor. The database includes server information from entirely different individual hospitals and buildings from different regions and divisions. Because there is a variety of patients littered with totally different diseases, it would generate a lot of knowledge in tending services every day. Therefore, there is a need for additional resources to effectively method the data. Instead of storing an equivalent on one computer, cloud computing is used to method the accumulated data. The benefit of storing cloud environment information is that data from completely different sources is stored in one location and doctor will be able to access it anywhere, anywhere. The storage must be efficient so that the results can be quickly retrieved. Since current awareness is enormous in nature with large volume, values, and versatility; therefore, ancient strategies for storage and retrieval would be scarce to system information with respect to parameters such as delay and production. Cloud computing is used as the basis of tending as-a-service (HaaS) in the planned process in managing healthcare services. The area unit of cloud services used to store the knowledge perceived by the patient's body in real time. This information is gathered through a variety of body sensors which area unit is deployed in the body of a person or in the kind of wearable sensors. Extending ancient storage usefulness, our research does not only provide doctors with fast and reliable data, but it can also predict the long-term illnesses a patient may experience. It needs to provide signs of a disease provided by an expert or a professional physician (Jindal et al., 2018).

8.2.4 MACHINE INTELLIGENCE IN THE DISCOVERY OF RATIONAL DRUG

AI means intelligence expressed in computers. To direct traditional studies that are costly and time consuming in the sense of rational drug discovery, numerous mechanical intelligence approaches have been used. A linear and nonlinear approach can be defined as a Quantitative structure–activity relationship (QSAR) tool used in drug discovery. For example, Belhumeur et al. (2018) introduced linear discriminant analysis (LDA) for pattern recognition and AI in 1996. LDA is a classifier who sees a linear equation as maximizing the distance between classes and minimizing the distance within class. LDA was used for predicting drug–drug interactions (Vilar et al., 2015a), identifying new compounds (Medina Marrero et al., 2015), and, among others, detecting adverse drug events (Vilar et al., 2015b). Although LDA is a simple approach, a powerful modeling method is still considered to be the combination of LDA and novel descriptors. For example, in conjunction with topological, 3D chiral, topographical, and geometric descriptors, Marrero et al. used an LDA algorithm to predict drug antifungal activity and provide greater accuracy compared to other nonlinear approaches. Vapnik and Vapnik (1998) have suggested support vector machines (SVMs). For their ability to handle small data sets of high-dimensional variables. Regarding linear problems the SVM model divides different categories by mapping spaces to optimize the distance between various dots classes (Poorinmohammad et al., 2015). SVMs use kernel mapping for nonlinear problems and convert nonlinear data sets into a high-end function space for linear classification. SVM was commonly used in drug discovery for different modeling purposes (Jain et al., 2015). Poorinmohammad et al. combined the SVM method with descriptors to classify 96.76% predictive accuracy anti-HIV pseudo-amino acid composition. DTs are a simple and interpretable framework for ML. Two significant steps are generally required in the development of DTs: the collection and pruning of attributes. Initially, the attributes of a molecule are chosen as a test (e.g., whether the partition coefficient of the molecule is greater than 5). The selected attributes are treated as internal nodes, including the root node and the nonleaf nodes. The branch is the result of the test and the leaf node is a grading label. Second, tailoring algorithms are used to decrease the complexity of the list. DTs recently have been utilized in modeling the properties of the absorption, distribution, and metabolism and toxicity of drugs (Newby et al., 2015; Gupta et al., 2015). The gradient-decending approach which reduces the medium-square error between the network output and the experimental training set data. Back-Propagation Neural Networks (BPNN) features robotic performance, superior tolerance for errors, parallel coprocessing, self-organization, and self-learning. Not only was BPNN used in QSAR (Zhang et al., 2015).

8.2.5 APPLICATION OF MACHINE LEARNING IN DISEASE RISK PREDICTION AND PROGNOSIS

Before studying to know the best option of choosing ML algorithms for a particular problem, we should have a proper understanding about what ML is and what it is not about. ML is the challenging research area having wide applications under AI, which utilizes past experience (captured examples by itself) to learn itself. It uses a different set of training/learning tools related to such as research on statistics, theory of probabilistic analysis, and optimization techniques and once it achieves learning capability at significant level, it is ready to predict and classify new input. Now this model is ready to recognize new patterns by predicting current novel trends (Mitchell, 1997).

Some of these methods, like statistics-based learning can be applied on the problem dataset to extract useful pattern. In addition to statistical parameters, these methods utilize other parameters also to make more correct learning and decision. These parameters are Boolean logic parameters (AND, OR, NOT), conditional probabilistic parameters, conditional-based restrictions (parameters: IF, THEN, ELSE), and also parameters of other nontraditional optimization methods. Nowadays, these methods are widely used for modeling the real-world data or in applications which require extracting useful pattern information for better classification. The approach used by these techniques resembles human. Although most of the ML methods incorporate the techniques which are majorly derived from statistics and probability-based theory, eventually nowadays they have become more powerful tool for classification task. These techniques are good enough to make correct decisions or inference from even complex dataset, whereas traditional techniques are not able to achieve this milestone (Duda et al., 2001).

Often, the performance of the statistical methods is better in combination with multivariate regression or correlation analysis. But they have drawback of use preassuming the variable independence. These techniques also combine the variables linearly to build better classification model. Therefore, in contrast to traditional statistical techniques being stuck into local optimum solution, these techniques work well in case of different relationships among variables such as nonlinear, interdependence, or conditional dependence. They can be applied on systems having inherent features similar to complex biological models. Only a few of systems like simple physical models are linear in nature and possess dependent variables. Hence, ML-based techniques have emerged as better classification methods to get better performance.

MLTs always do not give guarantee to yield optimum results. The success of any problem-solving technique heavily depends on how deeply we understand the given problem and also the underlined limitations such as limited availability of required datasets for analysis. Therefore, well-ordered assumptions and underlined limitations may lead to get improved results. The proper design of experimental setup of these techniques may produce better performance. This success also depends on how better we tune on learning tool (training tool) and validation of classification with variety of features of dataset. The well-known fact of research done till now that poor-quality data merely produce poor-quality results (similar to garbage-in is equal to garbage-out). The advent of such deep-learning models which has drawback of having more variables than number of events to be predicted, fail to perform better regardless of database heterogeneity. This problem is known as "curse of dimensionality," not only limited to the scope of ML applications but also traditional techniques such as statistical classification methods. Therefore, a need arises such that either to reduce the size of variables (feature-set) or training dataset should have more and more examples. The research done by Somorjai et al. (2003) proposed that the ratio of sample data-per-feature unit should be more than 5:1 for better training and classification. The success of these techniques not only determined by the volume of training/learning examples but also the presence of variety of data in training dataset (should have different possible examples with corresponding input and output data). The programmer build the training examples by considering large portions of relevant data which is powerful to perform dimensional reduction of the dataset as per learning tool expectations. These methods have drawback of overtraining or training with noisy concept, because of availability of only few examples with less variety of inherent features (Rodvold et al., 2001). The behavior of the overtrained model is just like an overtired student which poorly performs in writing exams. It also produces poor results in classifying and recognizing the novel input data.

In general, ML methods employ three types of learning approaches: (i) supervised learning; (ii) unsupervised learning, and (iii) reinforcement learning. The selection of these learning methods is done based on the desired outcome of the given classification problem which need to employ ML algorithm (Duda et al., 2001). The supervised learning methods have a prerequisite of a teacher who provides a large set of examples to train the model. This method tries to extract relationship with higher degree between labeled input and output data among training dataset.

For example, as shown in Figure 8.4, the labeled training data have lableled input as corrupted images of numeric symbol "8," and corresponding output image is the correct image of the same number. The learning unit learn from experience like human from the data in the training dataset. The trainer helps in learning the pattern what is supposed to find out as outcome, similar to how most of the school students used to learn. In contrary, the unsupervised learning techniques usually work on unlabeled dataset (no mapping of input to output) which follows the pattern similar to the learning process of graduate students. The most widely used unsupervised learning techniques are K-means clustering, Hierarchical clustering, self-organizing feature maps (SOMs) and other clustering methods, and so on. The approach used by these algorithms is by extracting inherent features from unlabeled or unclassified dataset to construct groups or clusters helps in building classification model such as disease prescreening tool.

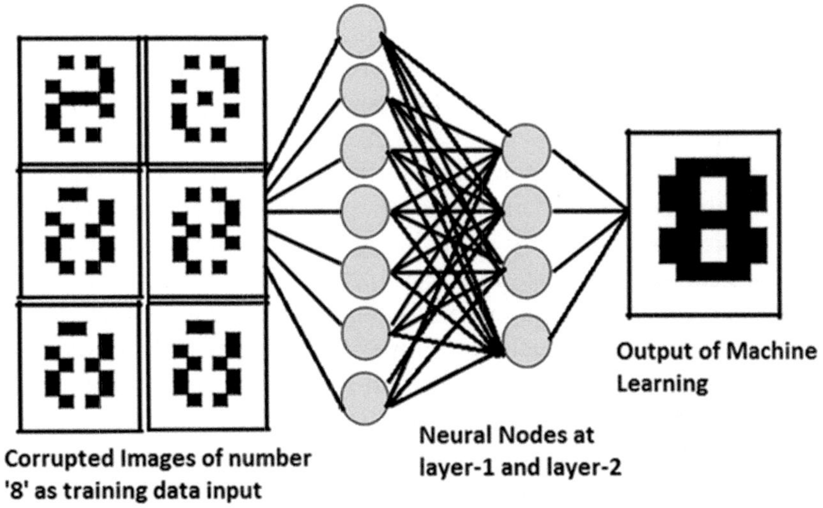

FIGURE 8.4 Training model architecture to recognize numeric symbol "8" using input training set (a set of corrupted images of numeric symbol "8") to map with correct output image of the same.

The learning methodology used by SOM-based model (Kohonen, 1982) is a special example of an artificial neural network (ANN) classifier. In this method, the individual weight vectors of artificial neuron-based grid of training and learning network are updated to find out the similarity and dissimilarity feature from input data for successful training. The SOM-based classification model originated from the concept of biological neural-brain function which inherently forms a set of artificial neurons. Each neuron unit is placed on its own location on the output map of SOM architecture by searching method. All the processing units of SOM usually compete each other during winner-take-all process in its competitive neural network. In this process, a processing unit is declared as a winner node which has weight vector closest to the input. Thereafter, the winner node weights are updated in adaptive way such that it becomes more close to the corresponding input vector. For the purpose of fine-tuning, the weights of neighboring nodes, which are closely connected to winning node are updated with degree of high variation. On the contrary, the weights of distant neighboring nodes are fine-tuned with smaller variation. This complete process is repeated for all the training input vectors for a number of iterations individually. At the last step, we get a set of winning nodes with appropriate weights a corresponding set of input vectors. Finally, a self-organizing map model corresponding to given problem is constructed as a neural-net result. This model has resultant associations (connective weights) in between each parts of output nodes with corresponding input data in the dataset.

Apart from classification models using unsupervised learning, nowadays analytical researchers are using supervised learning method for ML applications especially in the implementation of diagnostic systems for medical analysis such as disease prescreening, risk prediction, and further process of disease prognosis. These learning algorithms rely on the principle of conditional probabilities to make decisions or predictions usually followed by most of the current-age classifiers. These algorithms are mainly categorized into (i) ANN, (ii) genetic algorithm (GA), (iii) k-nearest neighbor algorithms (k-NN), (iv) SVM, (v) DT, (vi) LDA methods, and so on. Most of the analytical researchers have used ANN as a conditional algorithm in the classification models which can be employed in variety of domains.

A large set of pattern recognition and classification problems of many real-time applications can be handled by ANN-based conditional algorithms. These algorithms can perform a range of operations which are part of a classification process. Some of these are statistical operations such as linear, logistic, nonlinear regression analysis, and logical operations such as AND, OR, NOT, XOR, IF–THEN. The result of utilizing these features build more robust classifiers that they are efficient enough to solve a range of variety of classification problems (Rodvold et al., 2001). The idea of designing the architecture of ANN is especially followed from the human-brain network. It has millions of neurons interconnected and their junctions act as axons in human-brain network. Its designing is to mimic the functionality of the human brain by forming mathematical modeling corresponding to the neurophysiological structure. As human learns, the connection strength (weights and biases) of neuron connections of ANN is updated by providing a variety of training examples through a number of iterations. In general, the authors save the strength of neuron interconnections in the form of a matrix, called weight matrix. As in Figure 8.4, the ANN classification architecture has multiple layers called hidden layers. These hidden layers help in promoting multilayered learning to process their inputs to generate more accurate outputs. While in the case of implementation, for learning stage, input and output data are mathematically transformed into vectors, or strings or set of numbers in such a way that completely represent the structure of individual layer of neural network. It is an important part of classification process where we need mapping of input/output data (such as a physical and statistical features, inherent characteristics, data in the form of image or signal, a disease prognosis, a list of diseases, etc.) to corresponding numeric vector. This task seems to be a big challenge in applications using neural networks. A well-known back-propagation optimization algorithm can be used to update/modify the strengths (interconnections–weights) of neural network architecture. In this method, the error (difference between actual output and target output) is backpropagated to network to minimize itself by updating weights in each iteration. In this method, the derivative-based gradient descent approach is used to reduce the error by comparing the corresponding output of one layer of the network to its foregoing layer. In other words, the assigned weights to each connections from weight matrices correspond to backpropogation-based ANN architecture are repeatedly updated to minimize the error. Generally, a learning method based on differentiable transfer function (similar to Sigmoid curve) is used in the background of backpropagation algorithm. For each different application, we need different design and architecture of ANN classification model and further it can be reformulated, customized, or optimized as per our newly enerated requirements. These models may sometimes subjected to poor performance or even sometimes lethargic approach of training/learning provided they have generic structure of the input/output scheme. Another shortcoming of backpropogation-based ANNs is that it resembles with "Black-box model" technology. It looks almost impossible to discern the approach how the classification task is performed in this model or why it is unable to perform for a particular case. In some other words, it is very difficult to decipher the background logic involved in learning and classification for a trained network of an ANN.

Although some other conditional algorithm like DT, which can be easily discerned in contrast to ANN's logic of DTs. Simply, the DT structure follows a graph or tree or flow chart structure in which nodes depict decisions and branches show their possible effects (Quinlan, 1986). DTs have been used for past years and are commonly used as a basic element of the classification or categorization phase for many systems based on medical diagnosis, like prescreening of diseases, that is, the breast cancer diagnosis process can be shown as a DT with node decisions and left as their effects, as depicted in Figure 8.5. In consultation with medical experts, the layout of DTs is constructed and thus modified or tuned through years of clinical practice experience. It can also be adjusted according to asset or risk limitations. On the contrast, DTs can also be built automatically by DT learners using labeled training dataset if they exist in real time. When classifying the dataset using a DT, tree's leaves represent classifications and branches show collections of feature vectors referring to correct classifications. After numerical or logical testing, the learning method of a DT is achieved by gradually splitting the labeled training dataset into subsets and this splitting process is repeated recursively on each freshly generated subset until further splitting is either a singular classification or no splitting is feasible. While DTs have several benefits over other conditional methods, they require few data preparation, they are simple to understand and define, and they can handle a wide range of data types such as numerical, nominal, and categorical data. They can also produce robust classifiers, learn quickly, and use statistical tests to validate them. In addition, DTs have bad performance similar to ANNs for a more complicated classification problem (Atlas et al., 1990).

Nowadays, an SVM has appeared as a new technique of ML (Duda et al., 2001). However, they are a well-known technique of ML, there is still more scope available in the field of prescreening illness, prediction of risk, diagnosis, and prognosis. Using the specified scatter plot of points of a specific problem, the working theory of SVM can be better understood. Scatter-plot (prescreening question for breast cancer) between tumor masses and amount of axillary metastases in two categories: poor prognosis and excellent prognosis as in Figure 8.6. Scatter plot has, of course, formed two distinguishable clusters, and SVM machine learners can find out the equation of a line that attempts to divide these two clusters to the maximum extent. In addition, if the scatter plot is taken between several than two variables (such as volume, metastases, and estrogen receptor content), after which SVM will have a planer separation line and if a number of variables are increased the separation will be described as a hyperplane. This hyperplane is developed in two classes by a subset of the underlying points, named as support vectors.

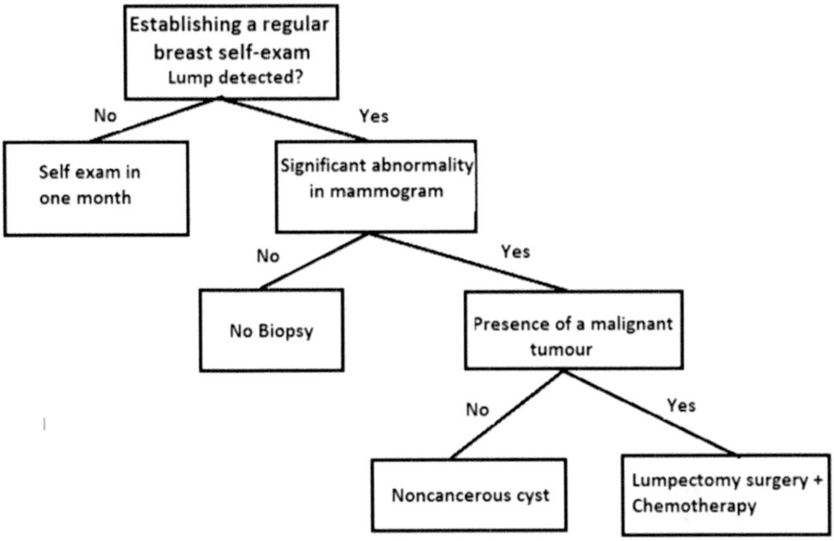

FIGURE 8.5 A decision tree-based flow of outcomes (used in prediction cases such as disease prescreening, diagnosis, and treatment of breast cancer). An expert assessment is followed to build tree.

Formally, a separating hyperplane is derived by the SVM algorithm. This plane classifies the dataset in such a way that it is partitioned into two classes comprising maximum margin. In other words, it tries to maximize the distance between the hyperplane and the closest examples taken from both the classes. Furthermore, it can handle the problems of nonlinear classification by applying nonlinear hybrid mathematical kernel functions. These kernel mathematical functions usually transform the data from a linear feature vector space (one-dimensional space) into a nonlinear feature vector space (other multidimensional space). It has been shown by researchers that the performance of an SVM classifier can be highly improved by applying different types of kernels functions in variety of domains. As similar to ANNs, SVM-based classifiers can handle a wide range of problems based on classification or pattern recognition such as text and speech recognition, protein function prediction, object segmentation, handwriting analysis, and prescreening tool for disease diagnosis and prognosis (Duda et al., 2001). Like k-NN-based classification, SVM training and learning has also better performance and proved to be very powerful for the problems especially, nonlinear classification problems (discussion shown in Table 8.2).

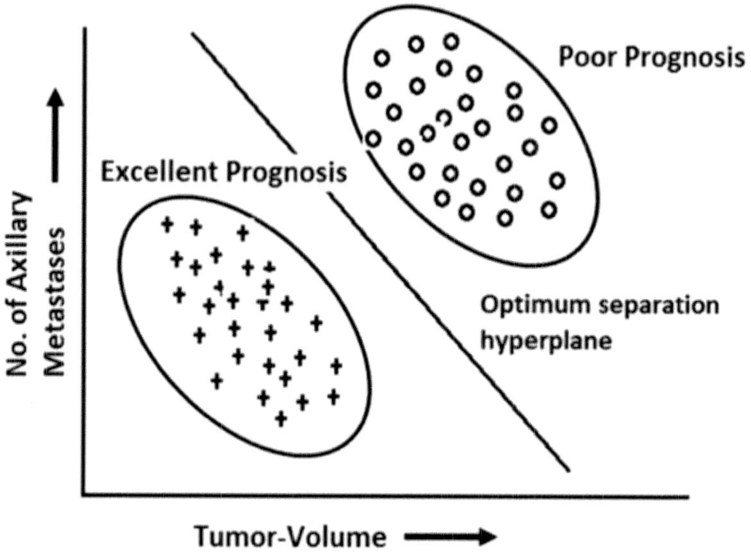

FIGURE 8.6 SVM-based classification approach between two entities: weightlifting players and basketball players considering their support vectors based on heights and weights. A hyperplane is derived by SVM classifier which increases the separation between the two different clusters up to maximum extent and act as actual separator.

8.2.6 APPLICATIONS OF MACHINE LEARNING IN BIOLOGY

8.2.6.1 IDENTIFYING GENE CODING REGIONS

Next-generation sequencing has become the fastest growing field of genomic science by sequencing a genome utilizing current age technology in a short period of time. Most of this software uses ML to classify gene coding regions within a genome. These gene prediction tools would be more adaptive and efficient with an ML method than traditional homologous sequence search techniques.

TABLE 8.2 Widely Used Examples of Machine Learning Techniques with Their Pros-Cons, Limitations, and Prior Assumptions to Develop Automated Diagnostic Systems in Medical Informatics.

S. No.	Machine learning techniques	Advantages in comparison to traditional techniques	Prior assumptions and disadvantages
1	Decision tree (Quinlan, 1986)	• Understandable and interpret an efficient method for training and learning • Need little preparation of data • Sequence of training and learning systems have no effect on training • Fast learning • The problem of over-fitting can be resolved with pruning	• Classes must follow the rule of mutual exclusion • The final decision tree depends on selection of attribute order • If the training set includes errors, excessively complicated decision trees would be formed • Testing of an attribute becomes vague when each branch of a decision tree is taken in case attribute has lost some of its values
2	k-Nearest neighbor (k-NN) (Patrick and Fischer, 1970)	• Quick classification of medical data • Robust and efficient algorithm towards novel approaches • Beneficial for complex and nonlinear problems used for classification • Work for both type of problems, that is, classification and regression analysis • Tolerate noisy instances or incomplete instances of attribute values	• Working slowly to update concept description • Assumes that data with similar attributes will be listed in a similar classification • Makes the assumption that all attributes have equal importance • As number of attributes increases, it also increases the computational complexity of algorithm
3	Naïve Bayes (Langley et al. 1992)	• Algorithm basically works on statistical modeling • Easy to understand and well-organized training and learning algorithm • Applicable for broad areas and domains	• Ensures statistical independence of features, regardless of whether they are present or not • Apply Gaussian distribution on various attributes or on numeric data • Redundant attributes can result in incorrect classification • Classes must work on the principle of mutual exclusion • Accuracy of algorithm is fully dependent on attribute and class frequencies

TABLE 8.2 *(Continued)*

S. No.	Machine learning techniques	Advantages in comparison to traditional techniques	Prior assumptions and disadvantages
4	Artificial neural network (ANN) (Rumelhart et al. 1986)	• Capable to manage a broad range of classification problems • Utilized for regression analysis or classification problem • Fully tolerate the data having noise • Boolean functions like AND, OR, NOT, etc. can be denoted by algorithm • Input data can be easily classified and shown by more than one output symbols	• Hard to understand the structure of algorithm and performance of the classification task • Many attributes may result in overfitting of the training architecture and may generate incorrect results in classification • Only repeated experiments can obtain optimal network structure
5	Support vector machine (SVM) (Vapnik, 1982)	• Capable to model boundaries of nonlinear class problem • Overfitting is unlikely to arise in SVM algorithm as compared to ANN • Reduce high computational complex problems to a quadratically optimize problems • Easily manage complicate decision principals and frequently occurring errors	• Slow training procedure as compared to decision trees and naïve Bayes • Optimal parameters are harder to find when training data are nonlinearly separable • Not easy to understand the structure of algorithm
6	Genetic algorithm (GA) (Holland 1975)	• An easy algorithm which is implemented based on selection, mutation, and crossover as follows the biological evolution • Basically used for feature selection and classification • Primarily working good in optimization problems • Find a good solution always which may lead to optimal solution but not the best solution	• Development of fitness function which is computationally non-trivial problems • Always tries to fetch local optima rather than global • Input/output data for training is hardly represented

8.2.6.2 STRUCTURE PREDICTION

Previously, the process of protein–protein interactions was used to predict protein structure. Nowadays, the application of ML in the prediction of protein structure has improved the accuracy from almost 70% to over 80%. Various application of ML approaches utilizing training sets in text mining has proved to be quite successful in finding novel drug targets from numerous literatures and looking for databases linked to secondary drugs.

8.2.7 DEEP LEARNING IN CLASSIFICATION OF CELLULAR IMAGES, GENOME ANALYSIS, AND DRUG DISCOVERY

Recently, deep learning is a specialized ML subdiscipline that is just the enhancement of neural network. The term "Deep" in deep learning typically refers to the number of layers that turn data into extensive information. Deep learning can therefore be considered identical to ANN containing multiple layers, that is, hidden layers. These multilayer nodes attempt to mimic how the human brain thinks to fix the problems. ML approaches are already using structure of neural networks. ML algorithms based on neural network technique need to perform analysis with refined or significant data from raw data sets. But data increases on genome sequencing made the processing of meaningful information very hard and therefore difficult to implement the analysis. Normally, multilayers in the neural network filter the information that is transmitted to each layer again and then allow the output to be refined by minimizing error. Deep learning algorithms retrieve appropriate features from large amounts of data such as dataset containing medical images or genomes and create a trained model based on training methods to extract features. Once the trained model has been developed, the respective algorithms can then use the model to analyze other data set. Currently, researchers related to medical field used deep-learning techniques to analyze and also figure out steps to relate genome information and medical image data to electronic health records. Today deep learning has become an important field of research in computational biology to construct complex systems for real-time processing. Usually, deep learning is applied to biological data containing high throughput which provide better understanding about complex data or high-dimensional dataset. For computational biology, for regulatory genomics, deep-learning approaches are used to classify regulatory variants, effect of mutation with the help of DNA sequencing, the population of cells and tissues, analyze the entire cells.

Although ML with artificial intelligence both are commonly used by various healthcare centers, healthcare professionals, and healthcare providers to increase patient satisfaction, provide more tailored services, improve efficiency and rational predictions, and thus improve the quality of human life. It is often used to create clinical trials more effective and to speed up the drug product development, testing and on time delivery.

8.3 CONCLUSION

For the past few years, the study of ML has placed remarkable milestones toward the automation of drug discovery, medical image analysis, genome analysis, prescreening tool for disease diagnosis and prognosis. These methods have successfully delivered significant improvements in performance as compared to other existing traditional algorithms. It has a wide range of applications for many real-life problems in day-to-day life, especially in biotechnology and medical science. Owing to their intensive performance in the development of classification and analytical systems, nowadays most of the researchers have framed a conception that within the next few years, ML-based real-time applications would take over human being in performance at par and the automated systems would be able help us in performing most of our daily routine tasks. However, application of ML in the field of healthcare and medical sector is limited to few numbers, especially in developing robust and more accurate medical diagnostic tools for drug discovery, medical image analysis, genome analysis, disease prescreening, and so on in comparison to the other existing traditional problems. In this chapter, variety of possible applications of MLTs in biotechnology and medical science were highlighted for the betterment of medical and healthcare industry. The obstacles faced by these techniques in these applications were also discussed that shows its limited applications employed yet in the field of healthcare sector. There arises a scope of utilizing these techniques in wide range for variety of applications in medical and biotechnology field. Although the list of discussed ML methods by no means complete in all the cases in medical science, it provides an indication of the wide range of impact of these techniques on the growth of the medical research industry. Finally, open issues and challenges faced by medical industry were also discussed to enrich the current research work, especially implementing the prescreening diagnostic tools.

ML-based applications are being used extensively by hospitals and healthcare-based service providers to improve patient satisfaction, deliver personalized treatments, make accurate predictions, and enhance the quality of life. It is also being used and proved to make clinical trials more efficiently and contributed to speed up the process of drug discovery and successful delivery. Although the era of ML is most promising nowadays to get AI-based development, these methods still have some limitations in applying over variety of domains. The availability of a plenty amount of high quality and complete data have great impact on the performance and reliability of ML-based automated system in variety of domains. First limitation for these methods is the availability of sufficient amount of biomedical data which is essentially required to emply. These datasets are generated by pharmaceutical companies which is normally not accessible in the public domain, but usually kept secret as expensive private assets by them. Second is, a lack of rational and correct interpretation of these data resembling with any of biological mechanisms is another limitation for these learning methods. Although these techniques have been proved to produce high prediction accuracies in many research work especially medical informatics, they still perform like a "black box model" that has limitation of unablility to disclose its equivalent biological mechanisms embedded within it required for modeling. In the era of big data-oriented new-age technology, ML has emerged as a potential research tool for drug discovery. As our database is growing large enough with fast collection of data and technology is also evolving, nowadays ML methods have become the most demanding tool for computer-aided drug design.

ACKNOWLEDGMENTS

We would like to give sincere thanks to those valuable authors whose references have been cited in this book chapter. We would like to gratefully acknowledge to my head of the department, Prof. S. P. Tripathi, Professor, CSE department, IET Lucknow, India, and my Ph.D. supervisor, Prof. Tapobrata Lahiri, Professor, IIIT Allahabad, India for continuous guidance to write down this book chapter and motivating us. Finally, we would like to give heartily thanks to all those contributors who were involved directly or indirectly in preparation of this work.

KEYWORDS

- **diagnosis**
- **prognosis**
- **deep neural networks**
- **convolutional neural network**
- **deep neural network**
- **deep learning**
- **decision tree**
- **k-nearest neighbor**

REFERENCES

Abdullah, R.; Sahibudin, S.; Alias, R. A.; Selamat, M. H. Collaborative Knowledge Management Systems for Learning Organisations. *J. Inf. Knowl. Manage.* **2005,** *4* (4), 237–245.

Atlas, L.; Cole, R.; Connor, J. T.; El-Sharkawi, M. A.; Marks, R. J., II; Muthusamy, Y.; Barnard, E. Performance Comparisons between Back Propagation Networks and Classification Trees on Three Real World Applications. *Adv. Neural Inf. Process. Syst.* **1990,** *2*, 622–629.

Ayers, R. *Artificial Intelligence (AI)—The Next Step in Biotech: Big Data & Technology*; 2018. https://www.experfy.com/blog/artificial-intelligence-ai-the-next-step-in-biotech (accessed March 19, 2018).

Belhumeur, P. N.; Hespanha, J. P.; Kriegman, D. J. Eigenfaces vs Fisherfaces: Recognition Using Class Specific Linear Projection. *Eur. Conf. Comput. Vis.* **2018,** *19*, 45–58.

Bengio, Y.; Goodfellow, I. J.; Courville A. *Deep Learning*, 2015.

Bouhriz, M.; Chaoui, H. Big Data Privacy in Healthcare Moroccan. *Elsevier Proc. Comput. Sci.* **2015,** *63*, 575–580.

Davenport, T. H. *The Human Side of Big Data and High-Performance Analytics*. Harvard Business School, 2012.

Davenport, T. H. How Strategists Use Big Data to Support Internal Business Decisions, Discovery and Production. *Strat. Leadership* **2014,** *42* (4), 45–50.

Dimitrov, D. V. Medical Internet of Things and Big Data in Healthcare. *Healthcare Inform Res.* **2016,** *22* (3), 156–163.

Duda, R. O.; Hart, P. E.; Stork, D. G. *Pattern Classification*, 2nd ed.; Wiley: New York, 2001.

Gorla, N. Somers, T. M.; Wong, B. Organizational Impact of System Quality, Information Quality, and Service Quality, *J. Strat. Inf. Syst.* **2010,** *19* (3), 207–228.

Greene, C. S.; Troyanskaya, O. G. Chapter 2: Data-Driven View of Disease Biology. *PLoS Comput. Biol.* **2012,** *8* (12), e1002816.

Gupta, S.; Basant, N.; Singh, K. P. Estimating Sensory Irritation Potency of Volatile Organic Chemicals Using QSARs Based on Decision Tree Methods for Regulatory Purpose. *Ecotoxicology* **2015,** *24*, 873–886.

Halaweh, M.; El Massry, A. Conceptual Model for Successful Implementation of Big Data in Organizations. *J. Int. Technol. Inf. Manage.* **2015,** *24* (2), 21–29 .

Holland, J. H. *Adaptation in Natural and Artificial Systems*; University of Michigan Press: Ann Arbor, MI, 1975.

Hou, C.-K. Measuring the Impacts of the Integrating Information Systems on Decision-Making Performance and Organizational Performance: An Empirical Study of The Taiwan Semiconductor Industry. *Int. J. Technol., Policy Manage.* **2013,** *13* (1), 34–66.

Jain, N.; Gupta, S.; Saprec, N.; Sapre, N. S. In Silico De Novo Design of Novel NNRTIs: A Bio-Molecular Modelling Approach. *RSC Adv.* **2015,** *5*, 14814–14827.

Jaradat, N. J.; Khanfar, M. A.; Habash, M.; Taha, M. O. Combining Docking-Based Comparative Intermolecular Contacts Analysis and k-Nearest Neighbor Correlation for the Discovery of New Check Point Kinase 1 Inhibitors. *J. Comput. Aid. Mol. Des.* **2015,** *29*, 561–581.

Jindal, A.; Dua, A.; Kumar, N.; Das, A. K.; Vasilakos, A. V.; Rodrigues, J. J. P. C. Providing Healthcare-as-a-Service Using Fuzzy Rule-Based Big Data Analytics in Cloud Computing. *IEEE J. Biomed. Health Inform. Issue* **2018,** *99*, 1–14.

Kohonen, T. Self-organized Formation of Topologically Correct Featuremaps. *Biol. Cybernet.* **1982,** *43* (1), 59–69. doi:10.1007/BF00337288.

Krizhevsky, A.; Sutskever, I.; Hinton, G. E. ImageNet Classification with Deep Convolutional Neural Networks. *Adv. Neural Inf. Process. Syst.* **2012,** *60* (6), 1–9.

Kumar, N.; Chilamkurti, N.; Park, J. H. ALCA: Agent Learning-based Clustering Algorithm in Vehicular Ad Hoc Networks. *Pers. Ubiquit. Comput.* **2013,** *17* (8), 1683–1692.

Kumar, N.; Kaur; K.; Jindal, A.; Rodrigues, J. J. P. C. Providing Healthcare Services On-the-Fly Using Multi-player Cooperation Game Theory in Internet of Vehicles (IoV) Environment. *Dig. Commun. Netw.* **2015,** 1 (3), 191–203.

Kumari, P.; Natha, A.; Chaube, R. Identification of Human Drug Targets Using Machine-Learning Algorithms. *Comput. Biol. Med.* **2015,** *56*, 175–181.

Kung, L.; Jones-Farmer, A.; Wang, Y. Managing Big Data for Firm Performance: A Configurational Approach. In: *Twenty-first Americas Conference on Information Systems AMCIS* 2015; pp 1–9.

Lakulu, M. Z. M.; Abdullah, R.; Selamat, M. H.; Ibrahim, H.; Nor, M. Z. M. A Framework of Collaborative Knowledge Management System in Open Source Software Development Environment. *J. Comput. Inf Sci.* **2010,** *3* (1), 81.

Langley, P.; Iba, W.; Thompson, K. An Analysis of Bayesian Classifiers. In: *Proceedings of the Tenth National Conference on Artificial Intelligence*, 1992; pp 223–228.

LeCun, Y.; Bengio, Y.; Hinton, G. Deep Learning. *Nature* **2015,** *521* (7553), 436–444.

Lu, N.; Cheng, N.; Zhang, N.; Shen, X.; Mark, J. Connected Vehicles: Solutions and Challenges. *IEEE Internet Things J.* **2014,** 1 (4), 289–299.

Mamoshina, P.; Vieira, A.; Putin, E.; Zhavoronkov, A. Applications of Deep Learning in Biomedicine. *Mol. Pharm.* **2016,** *13*, 1445–1454. doi:10.1021/acs.molpharmaceut.5b00982.

Mancini, M. Exploiting Big Data for Improving Healthcare Services. *J. e-Learn. Knowl. Soc.* **2014,** 10 (2), 23–33.

Medina Marrero, R. Marrero-Ponce, Y.; Barigye, S. J.; Echeverría Díaz, Y.; Acevedo-Barrios, R.; Casañola-Martín, G. M., García Bernal, M.; Torrens, F.; Pérez-Giménez, F. QuBiLs-MAS Method in Early Drug Discovery and Rational Drug Identification of Antifungal Agents. *SAR QSAR Environ. Res.* **2015,** *26*, 943–958.

Meena, K.; Suruliandi, A. Local Binary Patterns and Its Variants for Face Recognition. In: *IEEE-International Conference on Recent Trends in Information Technology, ICRTIT 2011*; pp 782–786.

Merelli, I.; Pérez-Sánchez, H.; Gesing, S.; D'Agostino, D. Managing, Analysing, and Integrating Big Data in Medical Bioinformatics: Open Problems and Future Perspectives. *BioMed Res. Int. Vol.* **2014,** *134023*, 1–13.

Min, S.; Lee, B.; Yoon, S. Deep Learning in Bioinformatics. *Brief. Bioinf.* **2017,** *18* (5), 851–869. doi:10.1093/bib/bbw068.

Mistry, P.; Neagu, D. Trundle, P. R.; Vessey, J. D. Using Random Forest and Decision Tree Models for a New Vehicle Prediction Approach in Computational Toxicology. *Soft Comput.* **2016,** *20*, 2967–2979.

Mitchell, T. *Machine Learning*; McGraw Hill: New York, 1997.

Newby, D.; Freitas, A. A.; Ghafourian, T. Decision Trees to Characterise the Roles of Permeability and Solubility on the Prediction of Oral Absorption. *Eur. J. Med. Chem.* **2015,** *90*, 751–765.

Palanisamy, V.; Thirunavukarasu, R. Implications of Big Data Analytics in Developing Healthcare Frameworks—A Review. *J. King Saud Univ.—Comput. Inf. Sci.* **2019,** *31* (4), 1–11.

Patrick, E. A.; Fischer, F. P., III. A Generalized k-Nearest Neighbor Rule. *Inf. Cntrl.* **1970,** *16* (2), 128–152. doi:10.1016/S0019-9958(70)90081-1.

Petter, S.; DeLone, W.; McLean, E. *Measuring Information Systems Success: Models, Dimensions, Measures, and Interrelationships. Eur. J. Inf. Syst.* **2008,** *17* (3), 236–263.

Poorinmohammad, N.; Mohabatkar, H.; Behbahani, M.; Biria, D. Computational Prediction of Anti-HIV-1 Peptides and In Vitro Evaluation of Anti-HIV-1 Activity of HIV-1 P24-Derived Peptides. *J. Pept. Sci.* **2015,** *21*, 10–16.

Quinlan, J. R. Induction of Decision Trees. *Machine Learn.* **1986,** *1* (1), 81–106. doi:10.1007/BF00116251.

Riordon, J.; Sovilj, D.; Sanner, S.; Sinton, D.; Young, E. W. K. Deep Learning with Microfluidics for Biotechnology. *Trends Biotechnol.* **2019,** *37* (3), 310–324. doi:10.1016/j.tibtech.2018.08.005.

Rodvold, D. M.; McLeod, D. G.; Brandt, J. M.; Snow, P. B.; Murphy, G. P. Introduction to Artificial Neural Networks for Physicians: Taking the Lid Off the Blackbox. *Prostate* **2001,** *46* (1), 39–44. doi:10.1002/1097-0045(200101)46:1<39::AIDPROS1006> 3.0.CO;2-M PMID:11170130.

Rumelhart, D. E.; Hinton, G. E.; Williams, R. J. Learning Representations by Back-Propagating Errors. *Nature* **1986,** *323*, 533–536.

Sadat, R. M. N.; Rakib, A.; Salehin, M. M.; Afrin, N. Efficient Design of Local Binary Pattern for Image Retrieval. In: *ISCI 2011 IEEE Symposium on Computers and Informatics*, 2011; pp 510–514.

Brooks's Life Sciences Blog. Sample Science Block, Brooks Life Sciences. *Machine Learning and Artificial Intelligence: The Golden Age of Medical Research*. https://www.brookslifesciences.com/blog/machine-learning-and-artificial-intelligence-golden-age-medical-research (accessed August 9, 2017).

Siddiqa, A.; Targio-Hashem, I. A.; Yaqoob, I.; Marjani, M.; Shamshirband, S.; Gani, A.; Nasaruddin, F. A Survey of Big Data Management: Taxonomy and State-of-the-Art. *J Netw. Comput. Appl. Vol.* **2016,** *71*, 151–166.

Singh, S.; Urooj, S.; Lay Ekuakille, A. Rotational-Invariant Texture Analysis Using Radon and Polar Complex Exponential Transform. *Adv. Intell. Syst. Comput.* **2014,** *327*. 10.1007/978-3-319-11933-5_35.

Somorjai, R. L.; Dolenko, B.; Baumgartner, R. Class Prediction and Discovery Using Gene Microarray and Proteomics Mass Spectroscopy Data: Curses, Caveats, Cautions. *Bioinformatics (Oxf., Engl.)* **2003,** *19* (12), 1484–1491. doi:10.1093/bioinformatics/btg182 PMID:12912828.

Surakarin, W.; Chongstitvatana, P. Classification of Clothing with Weighted SURF and Local Binary Patterns. In: *International Computer Science and Engineering Conference (ICSEC)*, 2015; pp 1–4.

Talab, A. R. R.; Shakoor, M. H. Fabric Classification Using New Mapping of Local Binary Pattern. In: *2018 International Conference on Intelligent Systems and Computer Vision (ISCV)*, Fez, 2018; pp 1–4.

Talib, A. M.; Atan, R.; Abdullah, R.; Azmi Murad, M. A. Towards New Data Access Control Technique Based on Multi-agent System Architecture for Cloud Computing, Digital Information Processing and Communications. *Communications in Computer and Information Science*; Springer: Berlin-Heidelberg, 2011; Vol. 189, pp 268–279.

Vapnik, V. Estimation of Dependences Based on Empirical Data. Springer Verlag: New York, 1982.

Vapnik, V. N.; Vapnik, V. *Statistical Learning Theory*; Wiley: Hoboken, NJ, 1998.

Vilar, S.; Lorberbaum, T.; Hripcsak, G.; Tatonetti, N. P. Improving Detection of Arrhythmia Drug–Drug Interactions in Pharmacovigilance Data through the Implementation of Similarity-Based Modelling. *PLoS One* **2015a,** *10* (6), e0129974.

Vilar, S.; Tatonetti, N. P.; Hripcsak, G. 3D Pharmacophoric Similarity Improves Multiadverse Drug Event Identification in Pharmacovigilance. *Sci. Rep.* **2015b,** *5*, 8809.

Wang, Y.; Guo, Y.; Kuang, Q.; Pu, X.; Ji, Y.; Zhang, Z.; Li, M. A Comparative Study of Family-Specific Protein-Ligand Complex Affinity Prediction Based on Random Forest Approach. *J. Comput. Aid. Mol. Des.* **2015,** *29*, 349–360.

Weidlich, I. E.; Filippov, I. V.; Brown, J.; Kaushik-Basu, N.; Krishnan, R.; Nicklaus, M. C.; Thorpe, I. F. Inhibitors for the Hepatitis C Virus RNA Polymerase Explored by SAR with Advanced Machine Learning Methods. *Bioorg. Med. Chem.* **2013,** *21*, 3127–3137.

Wiart, C. *Etnopharmacology of Medicinal Plants*; Humana Press: New Jersey, 2006; pp 1–50.

Wixom, B. H.; Todd, P. A. A Theoretical Integration of User Satisfaction and Technology Acceptance, *Inf. Syst. Res.* **2005,** *16* (1), 85–102.

World Health Organization, Special Programme for Research, Training in Tropical Diseases, World Health Organization, Department of Control of Neglected Tropical Diseases, World Health Organization. *Epidemic, Pandemic Alert. Dengue: guidelines for diagnosis, treatment, prevention and control*. World Health Organization, 2009.

Wu, P.-Y.; Cheng, C.-W.; Kaddi, C. D.; Venugopalan, J.; Hoffman, R.; Wang, M. D. Omic and Electronic Health Record Big Data Analytics for Precision Medicine. *IEEE Trans. Biomed. Eng.* **2017,** *64*, 263–273.

Xiao, S.-S.; Wu, Y.-X. Rotation-Invariant Texture Analysis Using Radon and Fourier Transforms. *J. Phys.: Conf. Ser.* **2007,** *48*, 1459–1464.

Yuan, J.-H.; Huang, D.-S.; Zhu, H.-D.; Gan, Y. Completed Hybrid Local Binary Pattern for Texture Classification. In: *Proceedings of the International Joint Conference on Neural Networks*, 2014; pp 2050–2057.

Zeng, X. Illumination and Rotation Invariant Texture Representation. In: *16th IEEE International Conference on Image Processing (ICIP)*, 2009.

Zhang, T.; Nie, S.; Liu, B.; Yu, Y.; Zhang, Y.; Liu, J. Activity Prediction and Molecular Mechanism of Bovine Blood Derived Angiotensin I-Converting Enzyme Inhibitory Peptides. *PLoS One* **2015,** *10* (3), e0119598.

Zhang, L.; Tan, J.; Han, D.; Zhu, H. From Machine Learning to Deep Learning: Progress in Machine Intelligence for Rational Drug Discovery. *Drug Discov. Today* **2017,** 22 (11), 1680–1685.

Index

D

E

G

N

P

W

X